Field Guide to Soils and the Environment

Environmental Resource Management Series

Consulting Editor

Donald R. Coates
State University of New York, Binghamton

Field Guide to Soils and the Environment
Applications of Soil Surveys

Gerald W. Olson

A Dowden & Culver book

Chapman and Hall
New York   London

First published 1984 by
Chapman and Hall
733 Third Avenue, New York NY 10017

Published in Great Britain by
Chapman and Hall Ltd
11 New Fetter Lane London EC4P 4EE

© 1984 Dowden & Culver, Inc.

Printed in the United States of America

ISBN 0 412 25960 5 (cased edition)
ISBN 0 412 25970 2 (paperback edition)

Library of Congress Cataloging in Publication Data

Olson, Gerald W.
  Field guide to soils and the environment.

  "A Dowden and Culver book."
  Bibliography: p.
  Includes index.
  1. Soil surveys. 2. Soil science. I. Title.
S592.14.O36 1984   631.4   84-5050
ISBN 0-412-25960-5
ISBN 0-412-25970-2 (pbk.)

This book is dedicated to my wife, Mary, who helped me to understand the humanistic aspects of soil survey interpretations and land uses. As an Ordained United Methodist Minister, she sometimes asks me to give sermons in her churches on such occasions as World Food Day and Freedom from Hunger campaigns. Her countless hours spent in distribution of food to the poor have accentuated to me the importance of reordering priorities for resource use and improving planning and management schemes for soils and land uses in the future.

Foreword

The success of the book *Soils and the Environment* illustrates the need for further, more detailed information about soil survey interpretations (uses of soil surveys), especially for laypersons, teachers, and students. Much information about soils and the environment is secluded in offices of various agencies and institutions and thus is not readily available to the people who need it. Techniques for finding and using the information are also not well known, so there is great need for this *Field Guide to Soils and the Environment* to provide teachers and learners with exercises that will give them practice leading to confidence in the manipulation and utilization of soil survey data. In a sense, all of us are (or should be) learners and teachers in the use of soil survey information. This *Field Guide* therefore emphasizes the use of initiative and imagination in the applications of soil surveys, toward the end of improving productivity and efficiency in the use of soils and the environment. Although laypersons, teachers, and students are the primary groups addressed by this *Field Guide*, other people involved with using soil surveys are (or will be) agriculturalists, agronomists, assessors, botanists, conservationists, contractors, ecologists, economists, engineers, extension workers, foresters, geologists, groundwater experts, planners, politicians, public health officials, range managers, recreationists, soil scientists, wildlife specialists, and many others. This *Field Guide* complements and enhances the book *Soils and the Environment* published in 1981.

DONALD R. COATES

Preface

Exercises in this *Field Guide to Soils and the Environment* are segregated into three parts:

Part I. Language of Soil Surveys and Criteria for Soil Ratings

Part II. Applications of Soil Surveys in Systems of Wide Usage

Part III. Principles Governing the Applications of Soil Survey Interpretations in the Future

Part I deals first with the basic data of the soil survey: namely, the soil profile descriptions, soil maps, and laboratory analyses. The process of classification is addressed through exercises on Soil Taxonomy and computerized groupings of soils. Practical work on projects is introduced early in the *Field Guide* (as in sequence in course teaching), so that students can have the maximum amount of time to work on topics of individual interest as the course progresses. Photographs are emphasized as one of the best tools to describe and teach about soils and land uses. The first exam provides a "learning experience" for students to evaluate the basic data, arrange soils into useful categorical groups, and make astute judgments and predictions about soil behavior.

Part II provides information and experience in applications of soil surveys in systems of wide usage. Initial emphasis on engineering applications and waste disposal illustrate how societal (and individual) problems can be solved by using soil surveys. Rural programs are described and experiences provided in agricultural land classification, erosion control, yield correlations, and farm planning. An exercise on community planning shows how the total environment should be considered for most efficient and effective uses of resources in urban as well as rural areas.

Part III covers the principles governing the applications of soil survey interpretations in the future. The exercise on soil potentials provides experience in evaluating inputs necessary to achieve maximum returns from each specific soil. Soil variability evaluations help students to appreciate and describe the "range in characteristics" of soil map units, and express the variability in meaningful terms. Sequential testing offers a tool by which sequences of soils in landscapes can be evaluated

by different yields and other production and performance functions. The exercise of land uses and soils makes correlations that enable understanding of effects of specific soil characteristics. Exercises of the tragedy of the commons, strategic implications, and military campaigns emphasize that soils are resources of extreme value, and that their uses and abuses help to determine the fate of nations—ultimately even in warfare. Short sections on research and predictions open avenues for future work—to challenge laypersons, teachers, and students. Soils tours and slide sets are emphasized as effective teaching devices. Formats for a final exam and an evaluation of the teaching are provided.

This *Field Guide* is intended to accompany the textbook *Soils and the Environment* * for use in teaching and learning about the applications of soil surveys. The *Field Guide* exercises give a series of "hands-on" experiences to students that will take them out into the field, into the laboratory, into cartographic manipulations, and into contact with the people making and using soil surveys. The perspective of all the exercises is that of soil survey interpretations (uses of soil surveys). The format is flexible so that all or part of the exercises can be used in a full course or short courses in formal or informal settings.

ACKNOWLEDGMENTS

This *Field Guide* is a culmination of ideas and concepts accumulated over more than 25 years. The author is particularly grateful to A. Milo Dowden, Publisher, for suggesting the initial idea for a *Field Guide to Soils and the Environment*, and to Barbara Zeiders for editing the manuscript. Don Coates of the State University of New York has been especially supportive in the application of soil surveys and geomorphology information to en-

*Olson, G. W. 1981. Soils and the environment: A guide to soil surveys and their applications. A Dowden and Culver Book. Chapman and Hall, New York, London. 178 pages. Available from Methuen, Inc., 733 Third Avenue, New York, NY 10017. $29.50 hardcover. $16.95 paperback.

vironmental matters. Many materials in the *Field Guide* have come from the author's colleagues, especially the National Soils Handbook and publications of the Soil Conservation Service of the U.S. Department of Agriculture. Particularly fruitful were discussions during sabbatical work with the Soil Conservation Service in Washington, D.C., with Gerald Darby, Donald McCormack, and Keith Young. Students at Cornell University in Agronomy 566, "Use of Soil Information and Maps as Resource Inventories," and other courses the author has taught have provided inspiration and guidance to the exercises presented in this *Field Guide*. The author's work in extension and service as President, Secretary, and Treasurer of the Empire State Chapter of the Soil Conservation Society of America through the years has provided working relationships with most of the "soils people" in New York State. Fred Gilbert, Will Hanna, John Warner, and Dick Babcock have been particularly helpful. Many extension workers in New York State have provided "real-world" problems and solutions relating to soils and the environment appearing in this *Field Guide*. Colleagues at Cornell University and in other places have also contributed greatly to this work. Sandra Seymour typed the manuscript, and Eileen Callinan drafted the illustrations.

To the lay learner

This *Field Guide* is designed primarily for the layperson who does not have formal training in soils. It is intended to provide additional details and learning exercises to supplement the textbook *Soils and the Environment.* Each person or class can select the parts of the subject matter and the exercises that are relevant to appropriate aspects of soil survey interpretations—to help solve soils problems and to obtain training in the application of soil survey information. Everyone interested in soils can use the textbook and *Field Guide* as information sources to the uses of soil surveys. Locally, one can also seek help from Conservationists of the Soil Conservation Service, Cooperative Extension Agents, and others to get in touch with Soil Surveyors and Soil Scientists of the Cooperative Soil Survey. Soil Surveyors are a solitary lot, spending a great deal of time by themselves doing soil survey, and most are delighted to receive inquiries about soil surveys and their uses. Most colleges, universities, high schools, and elementary schools teach courses or parts of courses on soils and earth sciences, and this *Field Guide* and textbook information will be increasingly taught in the schools. The goal of the *Field Guide* and textbook is to provide basic information and details, and to inspire initiative and imagination in the lay learner and others about uses of soil surveys. Soils are basic resources to be used for the benefit of all of us, and we should all know more about our soils in order to improve their uses and enhance our stewardship of them.

To the teacher

This *Field Guide* consists of a series of exercises to give you details and to serve as a guide in teaching about soil survey interpretations. The *Field Guide* and accompanying textbook, *Soils and the Environment*, are written for all potential users, so that each teacher must adapt the materials and exercises to fit the specific course or informal teachings for which he or she is responsible. The teaching must also be adapted to fit the soils of the local environment. Ideally, all references cited should be procured by each teacher, so that a reference library is available as a knowledge base for the teacher and for consultation by the students. Soil surveys available locally (and Soil Scientists working in nearby areas) should be utilized to the maximum extent possible. Photographs should be liberally used to illustrate soil- and land-use conditions to the students. Initiative and imagination is encouraged in teaching. The exam format, for example, should be available to the student beforehand as a learning experience, but each teacher should adapt specific questions to the local specific soils environment and the needs of the specific class. Each teacher should use initiative and imagination in selection of a local study area with highly contrasting soils within short distances. Students especially like fieldwork and "hands-on" experiences, so that teachers should emphasize the project and problem-solving orientation of the exercises. The basic importance of soils emphasizes to all of us that we should know more about them in order to manage them better in the future.

To the student

This *Field Guide* is a series of exercises to be adapted and modified by your teacher to supplement the textbook *Soils and the Environment.* Soil surveys are rapidly gaining prominence as resource inventories, and it is extremely important that soil survey interpretations be given more emphasis in the future. Many jobs and careers in the future in areas of natural resource management will be involved with soil surveys. Thus your study and preparations in this subject-matter area will be a good investment of your time. If the material at times seems overwhelming, do not be discouraged. Do not be hesitant in directing inquiries to your teacher and to others who regularly work with soil surveys in a professional capacity. Develop your talents in accessory areas such as photography. The project and exercise emphasis of this *Field Guide* encourages you to get out into the field, into the laboratory, and into efforts that have relevance toward improvement of your community. Although soil map units are complex entities in landscapes, they can be described, characterized, classified, and utilized in a systematic and efficient manner. Ultimately, your learning of this material is to prepare you to teach others. As you learn, keep the ultimate role of teaching the subject matter in your mind. The "multiplier effect" applies as "each one teaches one" and many more. Knowledge of our soils is vital for progress in our society, and soil survey interpretations will be of increasing importance to you and to others in the future.

Contents

Language of soil surveys and criteria for soil ratings

Soil profile descriptions

PURPOSE The purpose of this exercise is to take the student into the field to dig in soils, to appreciate the characteristics of soil differences outlined in pages 1–40 of the textbook. This exercise can consist of several short field trips, or several half-days of describing different soils according to the format given on pages 8 and 10 of the textbook.

PROCEDURE Dig several pits in the middle of several different, well-defined landscape segments, as illustrated in Figure 1. The pits should exhibit highly contrasting soil properties. Each student should describe the environment and horizon characteristics in each pit according to Tables 1 and 2. Preparation includes discussion of pages 1–40 of the textbook and instruction on soils of the local area. Classroom exhibits of sand, clay, mottles, blocky and platy structures, and so on, can provide useful preparation for fieldwork. Instructions in determination of textures, consistence, and pH can be conveniently done in the classroom before students go into the field to make a soil profile description. Pages 24–26 of the textbook describe a practical exercise for determining soil texture that has proved to be quite useful in preparing to teach about soil profile descriptions.

COMPETITION A competitive element can be introduced into the student activity by using the forms illustrated in Tables 3 and 4. Under the sponsorship of the American Society of Agronomy, soils contests are held each fall (regionally) and each spring (nationally) in the United States. Host states are rotated among colleges offering B.S. training in soils. Thus, by participating in soils contests for four years, undergraduates have the opportunity to see many different soils and to meet the people working in the profession. Of course, unofficial contests can be held among participants of any class or group. The scorecards in Tables 3 and 4 offer a convenient means to grade individual performance in description of a soil. Each scorecard is changed somewhat from year to year and modified to fit local soil situations. Typically, numerous pages of explanatory materials accompany the scorecards for the students and coaches to study. The scorecards also require some classifications and interpretive judgments to be made about the soils, so that the value of the descriptive data are readily apparent to the students.

CHANGES Soil survey is one area of study where change is a continual process and a mark of progress. In 1981, for example, soil profile horizon designations given on page 14 of the textbook were modified somewhat to fit the international nomenclature more closely. Guthrie and Witty (1982) provided a synoptic comparison between the old (1962) and new (1981) designations for soil horizons and layers, also given in Table 5. Of course, most current published soil survey reports contain the nomenclature given on page 14 of the textbook, and the new terminology will not appear in many soil survey reports until after a number of years have passed.

REFERENCE

Guthrie, R. L. and J. E. Witty. 1982. New designations for soil horizons and layers and the new *Soil Survey Manual*. Soil Science Society of America Journal 46:443–444.

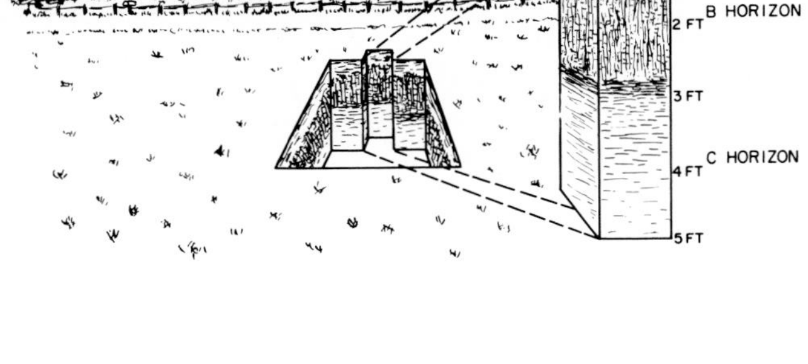

FIGURE 1/*Schematic diagram of a soil pit in a soil map landscape unit, from which the soil profile description area has been expanded to give a better view.*

TABLE 1/*Form for describing environment around the site (cut, pit, or trench) for a soil profile description.*

Soil type			File No.
Area		Date	Stop No.
Classification			
Location			
N. veg. (or crop)		Climate	
Parent material			
Physiography			
Relief	Drainage		Salt or alkali
Elevation	Gr. water		Stoniness
Slope	Moisture		
Aspect	Root distrib.		
Erosion			
Permeability			
Additional notes			

TABLE 2/*Form for describing horizons of a soil profile.*

Hori-zon	Depth	Color		Texture	Struc-ture	Consistence			Re-action	Bound-ary	
		Dry	Moist			Dry	Moist	Wet			

TABLE 3/Scorecard for the 1980 regional soils contest in New York State.

Contest Number _____

Pit Number _____

1980 NORTHEAST REGIONAL COLLEGIATE SOILS CONTEST SCORECARD
October 11, 1980
Cornell University

Score I _____
II _____
III _____
IV & V _____
TOTAL _____

Part I - Site Characteristics

POSITION OF SITE (5)
___ Floodplain
___ Stream terrace
___ Upland
___ Footslope
___ Depression

PARENT MATERIAL (5)
___ Residuum
___ Alluvium
___ Colluvium
___ Glacial Till
___ Loess
___ Marine and/or lacustrine
___ Glacial outwash

SLOPE (5)
___ Nearly level 0-3%
___ Gently sloping 3-8%
___ Moderately sloping 8-15%
___ Strongly sloping 15-25%
___ Moderately steep 25-35%
___ Steep 35+ %

EROSION (5)
___ Deposition
___ None to slight
___ Moderate
___ Severe

Part I - continued

SURFACE RUNOFF (5)
___ Ponded   ___ Medium
___ Very slow   ___ Rapid
___ Slow   ___ Very rapid

Part II - Soil Classification

EPIPEDON (5)
___ Anthropic   ___ Ochric
___ Histic   ___ Plaggen
___ Mollic   ___ Umbric

SUBSURFACE HORIZON (5)
___ Agric   ___ Natric
___ Albic   ___ Oxic
___ Argillic   ___ Petrocalcic
___ Calcic   ___ Petrogypsic
___ Cambic   ___ Placic
___ Duripan   ___ Salic
___ Fragipan   ___ Sombric
___ Gypsic   ___ Spodic
                ___ Sulfuric

ORDER (5)
___ Alfisol   ___ Mollisol
___ Aridisol   ___ Oxisol
___ Entisol   ___ Spodosol
___ Histosol   ___ Ultisol
___ Inceptisol ___ Vertisol

Part III - Soil Interpretation Limitations

SEPTIC TANK ABSORPTION FIELD (5)
___ Slight
___ Moderate
___ Severe

DWELLING WITH BASEMENT (5)
___ Slight
___ Moderate
___ Severe

LOCAL ROADS AND STREETS (5)
___ Slight
___ Moderate
___ Severe

LAND CAPABILITY CLASS AND
SUBCLASS (10)
___

TABLE 3 (continued)

Part IV - Profile description

Number of horizons to be described _____   Depth of horizons _____

Horizon (4)	Lower boundary depth, cm. (2)	Distinctness of lower boundary (2)	Texture (4)	Color (2)	Mottling		Structure		Moist consistence (2)	Score
					Abundance of mottling (1)	Contract of mottling (1)	Type (2)	Grade (2)		

Part V - Profile characteristics

INFILTRATION RATE (5)
_____ Slow
_____ Moderate
_____ Rapid
PERMEABILITY (5)
_____ Slow
_____ Moderate
_____ Rapid
EFFECTIVE SOIL DEPTH (5)
_____ Deep (>100 cm)
_____ Moderately deep (50-100 cm)
_____ Shallow (25-50 cm)
_____ Very shallow (<25 cm)

SOIL DRAINAGE CLASS (5)
_____ Somewhat excessively drained
_____ Well drained
_____ Moderately well drained
_____ Somewhat poorly drained
_____ Poorly drained
_____ Very poorly drained
AVAILABLE WATER-HOLDING CAPACITY (5)
_____ High
_____ Medium
_____ Low
_____ Very low

Score IV _____
V _____
Subtotal IV + V _____

TIEBREAKER
Horizon _____
_____ % Sand _____ % Silt _____ % Clay

6

TABLE 4/*Scorecard for the 1982 national soil contest in Arkansas.*

NATIONAL SOILS CONTEST: FAYETTEVILLE, ARKANSAS, APRIL 1982

CONTESTANT I.D.

SITE I.D.

PART I – SITE CHARACTERISTICS   PART III – SOIL CLASSIFICATION

POSITION OF SITE (5)   EPIPEDONS (5)   SUBSURFACE (5)

_____Flood Plain   _____Mollic   _____Argillic
_____Stream Terrace   _____Ochric   _____Cambic
_____Upland   _____Umbric   _____None
      _____No Epipedon   _____Fragipan
PARENT MATERIAL (5)   _____Lithic Contact
          _____Paralithic
_____Alluvium   Contact
_____Residuum

SLOPE (5)   ORDER (5)

_____0-1%   _____Alfisol
_____1-3%   _____Entisol
_____3-8%   _____Inceptisol
_____8-12%   _____Mollisol
_____12-20%   _____Ultisol
_____20% +

EROSION (5)   GREATGROUP (5)

_____Class 1   _____Fragiaqualf   _____Haplaquoll
_____Class 2   _____Glossaqualf   _____Paleudoll
_____Class 3   _____Albaqualf   _____Argiudoll
_____Class 4   _____Ochraqualf   _____Hapludoll
          _____Fragiudalf   _____Fragiaquult
          _____Paleudalf   _____Albaquult
SOIL LOSS (5)   _____Hapludalf   _____Paleaquult
t/ha·y   _____Fluvaquent   _____Ochraquult
          _____Udifluvent   _____Fragiudult
Bare_____   _____Haplaquept   _____Paleudult
          _____Eutrochrept   _____Rhodudult
Fescue_____   _____Dystrochrept   _____Hapludult
          _____Argiaquoll

Family Particle Size Class (5) _____ _____

FAMILY NAME: _____

(5 points for a correctly constructed name or 10 points for a name correct at all levels except the subgroup.)

TABLE 4 (continued)

PART II – SOIL MORPHOLOGY

Horizon (2)	Depth cm (2)	Boundary Dist. (2)	Sand % (1)	Clay % (1)	Texture (3)	Color (2)	Mottling (2)	Structure	
								Grade (2)	Shape (2)

PART IV – INTERPRETATIONS

HYDRAULIC COND./SURFACE (5)

_____ High
_____ Moderate
_____ Low

HYDRAULIC COND./SOIL (5)

_____ High
_____ Moderate
_____ Low

WATER RETENTION DIF. (5)

_____ High >22.5 cm
_____ Medium 15–22.5 cm
_____ Low 7.5–15 cm
_____ Very Low <7.5 cm

WETNESS CLASS (5)

_____ Class 1: >150 cm
_____ Class 2: 100–150 cm
_____ Class 3: 50–100 cm
_____ Class 4: 25–50 cm
_____ Class 5: <25 cm

USEFULNESS TO AGRICULTURE (8)

_____ Prime Farmland
_____ Additional Farmland of Statewide Importance
_____ Neither of the Above

ONSITE WASTEWATER RENOVATION (8)

_____ Standard Filter Field
_____ Modified Standard/Pres. Dist.
_____ Neither of the Above

TABLE 5/*Old and new designations for soil horizons and layers (adapted from Guthrie and Witty, 1982).*

Master Horizons and Layers

Old	New
0	O
01	Oi, Oe
02	Oa, Oe
A	A
A1	A
A2	E
A3	AB or EB
AB	—
A&B	E/B
AC	AC
B	B
B1	BA or BE
B&A	B/E
B2	B or Bw
B3	BC or CB
C	C
R	R

Subordinate Distinctions within Master Horizons

Old	New	
—	a	Highly decomposed organic matter
b	b	Buried soil horizon
cn	c	Concretions or nodules
—	e	Intermediately decomposed organic matter
f	f	Frozen soil
g	g	Strong gleying
h	h	Illuvial accumulation of organic matter
—	i	Slightly decomposed organic matter
ca	k	Accumulation of carbonates
m	m	Strong cementation
sa	n	Accumulation of sodium
—	o	Residual accumulation of sesquioxides
p	p	Plowing or other disturbance
si	q	Accumulation of silica
r	r	Weathered or soft bedrock
ir	s	Illuvial accumulation of sesquioxides
t	t	Accumulation of clay
—	v	Plinthite
—	w	Color or structural B
x	x	Fragipan character
cs	y	Accumulation of gypsum
sa	z	Accumulation of salts

Soil maps

PURPOSE This exercise is intended to relate the preceding exercise on soil profile descriptions to the areal expression of landscape segment differences on soil maps. Making soil maps involves a great deal of work and requires considerable experience of soil scientists, but anyone can learn about soil maps in accord with the relative investment of time and effort. This exercise can involve several half-days or full days in mapping the soils of the same study area in which the pits for the previous exercise were located. All the information on pages 1–40 of the textbook is used in making and evaluating soil maps.

PROCEDURE View aerial photographs of the study area stereoscopically (Soil Survey Staff, 1962) and delineate (in pencil) the major landforms (hills, valleys, plains, swamps, mountains, etc.). Locate on the aerial photos the pits from the preceding exercise. Then go into the field with spade and auger and investigate the variability of soils within each landscape segment. Observe vegetation and other land-use patterns. Walk along transects and tabulate statistical data on the soils. For example, in several 100-meter transects (perpendicular to contour lines), take borings at 10-meter intervals to get a percentage estimate of soil variability. Identify areas different in soil color, texture, structure, consistence, pH, stoniness, depth to rock, relief, drainage, and so on. Do not attempt to show areas too small to plot conveniently on the map at the scale designated. Use symbols for wet spots, eroded knobs, rock outcrops, gumbo or scabby spots, and so on. Make soil separations that are significant to use and management for present or intended use. Then ink the map and put symbols within each soil boundary for each soil map unit. Write soil map unit descriptions for each soil, slope, and other delineation. Profile descriptions must be made from pits to represent the "modal" (central) concept of each soil. All the information should be organized into a soil survey report, including narrative, soil profile descriptions, map unit descriptions, and soil map of the area. In summary:

1. Delineate landform boundaries on air photo viewed stereoscopically.

2. Investigate soil variability within landforms by digging at regular intervals along transects.
3. Refine map legend and soil boundaries from soil profile descriptions and variability observations.
4. Make detailed soil map unit delineations within landforms by digging in the soils in accord with detail and scale of mapping.
5. Ink the map boundaries and symbols on air photos or base map.
6. Write soil map unit descriptions.
7. Integrate all the information into a soil survey report for the area.

EXAMPLES Figure 2 is an example of a detailed soil map made on an aerial photographic base. Only about 300 acres can be mapped in a day of hard fieldwork in this complex soils landscape. The original scale of publication was 1:20,000, so that areas of different soils as small as 1 hectare or several acres could be delineated. The soil legend for Tompkins County, for which Figure 2 is a part, is given in Table 6. The detailed legend for the detailed soil map identifies the specific soil name, texture, slope, erosion class, and depth to bedrock. Other soil characteristics are described in the text of the soil survey report (Neeley et al., 1965). Figure 3 is a reproduction of the soil map without the air photo base, which is particularly valuable for coloring or shading with different patterns and for making interpretations about uses of the various areas.

The same principles of soil mapping, of course, can be applied at any scale to any base map. Figure 4 is the central portion of the soil map of New York State, with the geologic and Soil Taxonomy (Soil Survey Staff, 1975) legend and a "limitation legend" (Cline and Marshall, 1977). The original scale of this map was 1:750,000, and each general soil map unit contains a variety of different soils— but in a predictable pattern in the landscapes. The general legend indicates the dominant glacial materials in which the soils have formed, and the dominant soils and general Soil Taxonomy categories. The limitation legend indicates that the primary problems are droughtiness, steep slopes, shallowness to bedrock, stoniness, and wetness.

Figure 5 illustrates another general soil map, for part of Thailand. The alluvial plain (delta) in which Bangkok is located stands out in the southwest (lower left) part of the map, and the uplands can be identified from the legend in Table 7. Although the map in Figure 5 is general and at small scale, the soil and geomorphic differences shown on the map are very great and of extreme significance for national and regional planning. Figure 6 illustrates the situation in the alluvial plain north of Bangkok, where vast areas are flooded during the monsoon season. Figure 7, in contrast, shows the soil condition typical of many of the uplands, where deforestation has recently taken place and where erosion is severe. Soil classification units in Table 7 can be roughly converted (updated) into Soil Taxonomy units accordingly to the following correlations (Soil Survey Staff, 1975):

Classification Units (Moormann and Rojanasoonthon, 1972)	Soil Taxonomy (Soil Survey Staff, 1975)
Regosols	Entisols
Alluvial soils	Inceptisols
Peat and Muck soils	Histosols
Low-Humic Gley soils	Alfisols
Non-Calcic Brown soils	Alfisols
Gray Podzolic soils	Alfisols
Red-Yellow Podzolic soils	Ultisols
Grumusols	Vertisols
Rendzinas	Mollisols
Brown Forest soils	Inceptisols
Red-Brown Earths	Alfisols
Reddish-Brown Lateritic	Oxisols
Reddish-Brown Latosols	Oxisols
Red-Yellow Latosols	Oxisols

REFERENCES

Cline, M. G. and R. L. Marshall. 1977. Soils of New York landscapes. Information Bulletin 119, New York State College of Agriculture and Life Sciences, Cornell University, Ithaca, NY. 62 pages and map.

Moormann, F. R. and S. Rojanasoonthon. 1972. The soils of the Kingdom of Thailand. Report SSR-72A, Ministry of Agriculture and Cooperatives, Dept. of Land Development, and Food and Agriculture Organization of the United Nations. 64 pages and maps.

Neeley, J. A., E. B. Giddings, and C. S. Pearson. 1965. Soil survey of Tompkins County, New York. U.S. Dept. of Agriculture, Soil Conservation Service, in cooperation with Cornell University Agricultural Experiment Station. U.S. Government Printing Office, Washington, DC. 241 pages and 38 soil map sheets.

Soil Survey Staff. 1962 (under revision). Soil survey manual. Agricultural Handbook 18, U.S. Dept. of Agriculture, U.S. Government Printing Office, Washington, DC. 503 pages.

Soil Survey Staff. 1975. Soil Taxonomy: A basic system of soil classification for making and interpreting soil surveys. Agriculture Handbook 436, U.S. Dept. of Agriculture, U.S. Government Printing Office, Washington, DC. 754 pages.

FIGURE 2/*Detailed soil map on aerial photograph of area southeast of Ithaca, New York (see page 2 of the textbook). Scale of original map 1:20,000 (Neeley et al., 1965).*

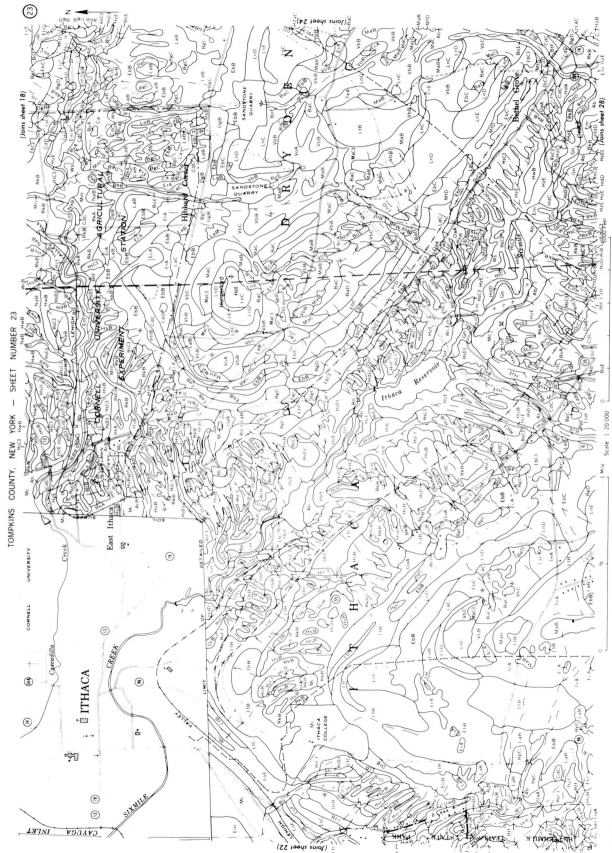

FIGURE 3/Detailed soil map without air photo background for area southeast of Ithaca, New York.

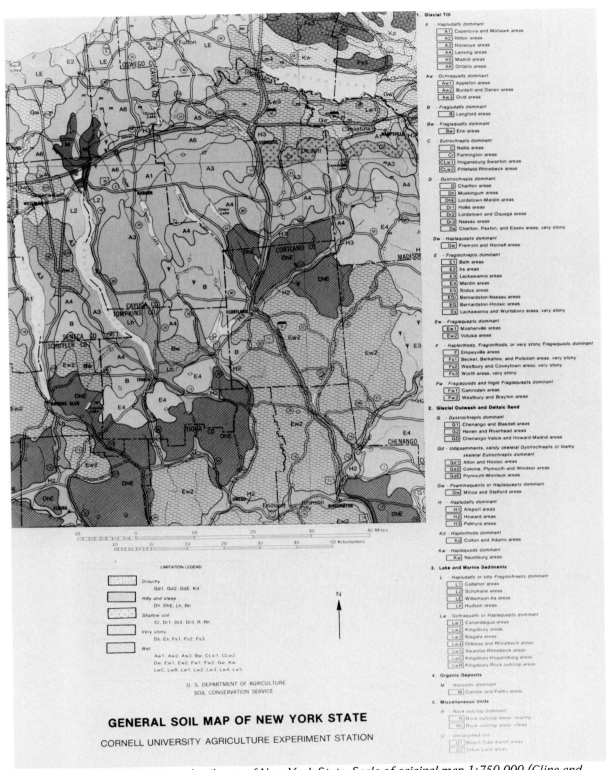

1. Glacial Till

A - Hapludalfs dominant
- A1 Cazenovia and Mohawk areas
- A2 Hilton areas
- A3 Honeoye areas
- A4 Lansing areas
- A5 Madrid areas
- A6 Ontario areas

Aw - Ochraqualfs dominant
- Aw1 Appleton areas
- Aw2 Burdett and Darien areas
- Aw3 Ovid areas

B - Fragiudalfs dominant
- B Langford areas

Bw - Fragiaqualfs dominant
- Bw Erie areas

C - Eutrochrepts dominant
- C Nellis areas
- Cr Farmington areas
- CLw1 Hogansburg-Swanton areas
- CLw2 Pittsfield-Rhinebeck areas

D - Dystrochrepts dominant
- D Charlton areas
- Dh Muskingum areas
- DhE Lordstown-Mardin areas
- Dr1 Hollis areas
- Dr2 Lordstown and Oquaga areas
- Dr3 Nassau areas
- Ds Charlton, Paxton, and Essex areas, very stony

Dw - Haplaquepts dominant
- Dw Fremont and Hornell areas

E - Fragiochrepts dominant
- E1 Bath areas
- E2 Ira areas
- E3 Lackawanna areas
- E4 Mardin areas
- E5 Sodus areas
- EDr Bernardston-Nassau areas
- EG Bernardston-Hoosic areas
- Es Lackawanna and Wurtsboro areas, very stony

Ew - Fragiaquepts dominant
- Ew1 Mosherville areas
- Ew2 Volusia areas

F - Haplorthods, Fragiorthods, or very stony Fragiaquods dominant
- F Empeyville areas
- Fs1 Becket, Berkshire, and Potsdam areas, very stony
- Fs2 Westbury and Coveytown areas, very stony
- Fs3 Worth areas, very stony

Fw - Fragiaquods and frigid Fragiaquepts dominant
- Fw1 Camroden areas
- Fw2 Westbury and Brayton areas

2. Glacial Outwash and Deltaic Sand

G - Dystrochrepts dominant
- G1 Chenango and Blasdell areas
- G2 Haven and Riverhead areas
- GO Chenango-Valois and Howard-Madrid areas

Gd - Udipsamments, sandy skeletal Dystrochrepts or loamy skeletal Eutrochrepts dominant
- Gd1 Alton and Hoosic areas
- Gd2 Colonie, Plymouth and Windsor areas
- GdE Plymouth-Montauk areas

Gw - Psammaquents or Haplaquepts dominant
- Gw Minoa and Stafford areas

H - Hapludalfs dominant
- H1 Arkport areas
- H2 Howard areas
- H3 Palmyra areas

Kd - Haplorthods dominant
- Kd Colton and Adams areas

Kw - Haplaquods dominant
- Kw Naumburg areas

3. Lake and Marine Sediments

L - Hapludalfs or silty Fragiochrepts dominant
- L1 Collamer areas
- L2 Schoharie areas
- LE Williamson-Ira areas
- Lh Hudson areas

Lw - Ochraqualfs or Haplaquepts dominant
- Lw1 Canandaigua areas
- Lw2 Kingsbury areas
- Lw3 Niagara areas
- Lw4 Odessa and Rhinebeck areas
- Lw5 Swanton-Rhinebeck areas
- LwC Kingsbury-Hogansburg areas
- LwR Kingsbury-Rock outcrop areas

4. Organic Deposits

M - Histosols dominant
- M Carlisle and Palms areas

5. Miscellaneous Units

R - Rock outcrop dominant
- R Rock outcrop areas - sloping
- Rh Rock outcrop areas - steep

U - Unclassified soil
- U1 Beach-Tidal marsh areas
- U2 Urban-Land areas

LIMITATION LEGEND

Droulhy
Gd1, Gd2, GdE, Kd

Hilly and steep
Dh, DhE, Lh, Rh

Shallow soil
Cr, Dr1, Dr2, Dr3, R, Rh

Very stony
Ds, Es, Fs1, Fs2, Fs3

Wet
Aw1, Aw2, Aw3, Bw, CLw1, CLw2,
Dw, Ew1, Ew2, Fw1, Fw2, Gw, Kw,
LwC, LwR, Lw1, Lw2, Lw3, Lw4, Lw5

N

U. S. DEPARTMENT OF AGRICULTURE
SOIL CONSERVATION SERVICE

GENERAL SOIL MAP OF NEW YORK STATE

CORNELL UNIVERSITY AGRICULTURE EXPERIMENT STATION

FIGURE 4/*Portion of the general soil map of New York State. Scale of original map 1:750,000 (Cline and Marshall, 1977).*

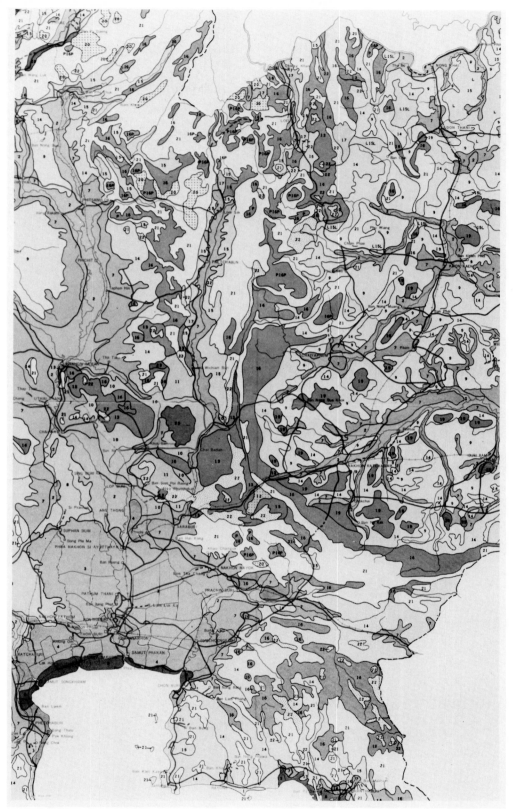

FIGURE 5/*General soil map of Thailand for area north-northeast of Bangkok. The Gulf of Thailand is in the lower left of the map, and the Laos border is at the top of the map to the north. Scale of original map 1:1,250,000 (Moormann and Rojanasoonthon, 1972).*

FIGURE 6/*View of alluvial plain north of Bangkok. During the monsoon season most of the plain is flooded with one to several meters of water, up to the bottom of the house built on stilts. A common crop of the area is "floating rice," which grows upward as the floodwaters rise, with the heads floating on the surface of the water. Often, the rice is harvested in boats.*

FIGURE 7/*Deforested and eroded hills northeast of Bangkok. Large areas have recently been cleared of forest in Thailand, and soil erosion is severe in many places. Soil maps could help a great deal to plan revegetation programs, and to manage watersheds so that environmental damage is minimized.*

TABLE 6/*Soil map unit legend for Tompkins County, New York.*

The first capital letter is the initial one of the soil name. A second capital letter, A, B, C, D, E, or F, shows the slope. A fourth letter, capital K in a symbol shows that the slope is complex. Some symbols without a slope letter are for nearly level soils or land types, but others are for soils or land types that have a considerable range in slope. A final number, 3, in the symbol shows that the soil is eroded.

SYMBOL	NAME	SYMBOL	NAME
Ab	Alluvial land	KaB	Kendaia silt loam, 3 to 8 percent slopes
ArB	Arkport fine sandy loam, 2 to 6 percent slopes	KnA	Kendaia and Lyons silt loams, 0 to 3 percent slopes
ArC	Arkport fine sandy loam, 6 to 12 percent slopes		
		LaB	Langford channery silt loam, 2 to 8 percent slopes
BaB	Bath channery silt loam, 2 to 5 percent slopes	LaB3	Langford channery silt loam, 3 to 8 percent slopes, eroded
BaC	Bath channery silt loam, 5 to 15 percent slopes	LaC	Langford channery silt loam, 8 to 15 percent slopes
BaC3	Bath channery silt loam, 5 to 15 percent slopes, eroded	LaC3	Langford channery silt loam, 8 to 15 percent slopes, eroded
BaD	Bath channery silt loam, 15 to 25 percent slopes	LbA	Lansing gravelly silt loam, 0 to 3 percent slopes
BgC	Bath and Valois gravelly silt loams, 5 to 15 percent slopes	LbB	Lansing gravelly silt loam, 3 to 8 percent slopes
BgC3	Bath and Valois gravelly silt loams, 5 to 15 percent slopes, eroded	LbB3	Lansing gravelly silt loam, 3 to 8 percent slopes, eroded
BgD	Bath and Valois gravelly silt loams, 15 to 25 percent slopes	LbC	Lansing gravelly silt loam, 8 to 15 percent slopes
BoE	Bath and Valois soils, 25 to 35 percent slopes	LbC3	Lansing gravelly silt loam, 8 to 15 percent slopes, eroded
BtF	Bath, Valois, and Lansing soils, 35 to 60 percent slopes	LmA	Lima silt loam, 0 to 3 percent slopes
BvA	Braceville gravelly silt loam, 0 to 5 percent slopes	LmB	Lima silt loam, 3 to 8 percent slopes
		LmB3	Lima silt loam, 3 to 8 percent slopes, eroded
Ca	Canandaigua and Lamson soils	LnC	Lordstown channery silt loam, 5 to 15 percent slopes
CdA	Chenango gravelly loam, 0 to 5 percent slopes	LnC3	Lordstown channery silt loam, 5 to 15 percent slopes, eroded
CdC	Chenango gravelly loam, 5 to 15 percent slopes	LnD	Lordstown channery silt loam, 15 to 25 percent slopes
CdD	Chenango gravelly loam, 15 to 25 percent slopes	LnE	Lordstown channery silt loam, 25 to 35 percent slopes
CfA	Conesus gravelly silt loam, 0 to 3 percent slopes	LoF	Lordstown soils, 35 to 70 percent slopes
CfB	Conesus gravelly silt loam, 3 to 8 percent slopes	LtB	Lordstown, Tuller, and Ovid soils, shallow and very shallow, 0 to 15 percent slopes
CfB3	Conesus gravelly silt loam, 3 to 8 percent slopes, eroded	LtC	Lordstown, Tuller, and Ovid soils, shallow and very shallow, 15 to 35 percent slopes
CnB	Chenango gravelly loam, fan, 0 to 8 percent slopes	Ly	Lyons silt loam
DgB	Darien gravelly silt loam, 2 to 8 percent slopes	MaB	Mardin channery silt loam, 2 to 8 percent slopes
		MaC	Mardin channery silt loam, 8 to 15 percent slopes
EbB	Erie channery silt loam, 3 to 8 percent slopes	MaC3	Mardin channery silt loam, 8 to 15 percent slopes, eroded
EbB3	Erie channery silt loam, 3 to 8 percent slopes, eroded	Mc	Made land
EbC	Erie channery silt loam, 8 to 15 percent slopes	MfD	Mardin and Langford soils, 15 to 25 percent slopes
EbC3	Erie channery silt loam, 8 to 15 percent slopes, eroded	Mm	Madalin mucky silty clay loam
EcA	Ellery, Chippewa, and Alden soils, 0 to 8 percent slopes	Mn	Madalin silty clay loam
Em	Eel silt loam	Mo	Middlebury and Tioga silt loams
ErA	Erie-Ellery channery silt loams, 0 to 3 percent slopes	Mp	Muck and Peat
FdB	Fredon silt loam, 0 to 5 percent slopes	NaB	Niagara silt loam, 2 to 6 percent slopes
Fm	Fresh water marsh		
		OaA	Ovid silt loam, 0 to 6 percent slopes
Gn	Genesee silt loam	OcC3	Ovid silty clay loam, 6 to 12 percent slopes, eroded
		OrA	Ovid and Rhinebeck silt loams, moderately deep, 0 to 2 percent slopes
Ha	Halsey silt loam	OrB	Ovid and Rhinebeck silt loams, moderately deep, 2 to 6 percent slopes
Hc	Halsey mucky silt loam	OrC	Ovid and Rhinebeck silt loams, moderately deep, 6 to 12 percent slopes
HdA	Howard gravelly loam, 0 to 5 percent slopes		
HdC	Howard gravelly loam, 5 to 15 percent simple slopes	PaA	Palmyra gravelly loam, 0 to 5 percent slopes
HdCK	Howard gravelly loam, 5 to 15 percent complex slopes	PaC	Palmyra gravelly loam, 5 to 15 percent simple slopes
HdD	Howard gravelly loam, 15 to 25 percent slopes	PaCK	Palmyra gravelly loam, 5 to 15 percent complex slopes
Hk	Holly and Papakating soils	PaD	Palmyra gravelly loam, 15 to 25 percent slopes
HmB	Honeoye gravelly silt loam, 2 to 8 percent slopes	PhA	Phelps gravelly silt loam, 0 to 3 percent slopes
HmC	Honeoye gravelly silt loam, 8 to 15 percent slopes	PhB	Phelps gravelly silt loam, 3 to 8 percent slopes
HmC3	Honeoye gravelly silt loam, 8 to 15 percent slopes, eroded		
HpE	Howard and Palmyra soils, 25 to 35 percent slopes	RhA	Red Hook gravelly silt loam, 0 to 5 percent slopes
HpF	Howard and Palmyra soils, 35 to 60 percent slopes	RkA	Rhinebeck silt loam, 0 to 2 percent slopes
HrC	Howard-Valois gravelly loams, 5 to 15 percent slopes	RkB	Rhinebeck silt loam, 2 to 6 percent slopes
HrD	Howard-Valois gravelly loams, 15 to 25 percent slopes	RnC3	Rhinebeck silty clay loam, 6 to 12 percent slopes, eroded
HsB	Hudson silty clay loam, 2 to 6 percent slopes	Ro	Rock outcrop
HsC3	Hudson silty clay loam, 6 to 12 percent slopes, eroded		
HsD3	Hudson silty clay loam, 12 to 20 percent slopes, eroded	TeA	Tuller channery silt loam, 0 to 6 percent slopes
HuB	Hudson-Cayuga silt loams, 2 to 8 percent slopes		
HuB3	Hudson-Cayuga silt loams, 2 to 6 percent slopes, eroded	VbB	Volusia channery silt loam, 3 to 8 percent slopes
HuC3	Hudson-Cayuga silt loams, 6 to 12 percent slopes, eroded	VbB3	Volusia channery silt loam, 3 to 8 percent slopes, eroded
HuD	Hudson-Cayuga silt loams, 12 to 20 percent slopes	VbC	Volusia channery silt loam, 8 to 15 percent slopes
HwB	Hudson and Collamer silt loams, 2 to 6 percent slopes	VbC3	Volusia channery silt loam, 8 to 15 percent slopes, eroded
HzE	Hudson and Dunkirk soils, 20 to 45 percent slopes	VoA	Volusia-Chippewa channery silt loams, 0 to 3 percent slopes
		VrD	Volusia and Erie soils, 15 to 25 percent slopes
IcA	Ilion silty clay loam, 0 to 2 percent slopes	Ws	Wayland and Sloan silt loams
IcB	Ilion silty clay loam, 2 to 6 percent slopes	WrB	Williamson very fine sandy loam, 2 to 6 percent slopes

TABLE 7/*Legend for the general soil map of Thailand (Moormann and Rojanasoonthon, 1972).*

I   EXCESSIVELY DRAINED SANDY SOILS, LOW FERTILITY

1   Regosols, on recent and semirecent beach and dune sand; level to undulating.

II   POORLY DRAINED CLAYEY SOILS, HIGH TO MODERATE FERTILITY

2   Alluvial soils, on recent fresh-water alluvium; level to gently undulating.
3   Alluvial soils, (acid sulphate soils) on recent brackish-water alluvium; level.
4   Alluvial soils, on recent marine alluvium; level.
5   Alluvial soils, saline, on recent marine alluvium; level.

III   VERY POORLY DRAINED ORGANIC SOILS (SWAMPS AND MARSHES)

6   Peat and Muck soils; level.

IV   POORLY DRAINED AND WELL DRAINED SOILS, MOSTLY LOAMY AND SANDY, MODERATE TO LOW FERTILITY

7   Low-Humic Gley soils, on semirecent and old alluvium; level to undulating.
8   Low-Humic Gley soils, and Noncalcic Brown soils, on semirecent alluvium, level to undulating.
9   Low-Humic Gley soils and Gray Podzolic soils, or Low-Humic Gley soils and Red-Yellow Podzolic soils with laterite, on old alluvium; level to undulating.

V   WELL DRAINED TO SOMEWHAT POORLY DRAINED LOAMY AND CLAYEY SOILS, MODERATE TO HIGH IN BASES, HIGH FERTILITY.

10   Grumusols and related dark-colored soils, on montmorillonitic clay from alluvium, marl and basalt; level to undulating.
11   Rendzinas and Brown Forest soils on marl and limestone alluvium; undulating.
12   Noncalcic Brown soils, on semirecent alluvium; level to undulating.
13   Red-Brown Earths, on residuum and colluvium from basic rocks (mainly limestone); undulating to rolling.

VI   WELL TO EXCESSIVELY DRAINED LOAMY AND SANDY SOILS, LOW IN WEATHERABLE MINERALS AND BASES, LOW FERTILITY.

14   Gray Podzolic soils, on old alluvium; level to undulating.

VII   WELL DRAINED CLAYEY AND LOAMY SOILS, LOW IN BASES, LOW FERTILITY

15   Red-Yellow Podzolic soils, on old alluvium; undulating to rolling.
L 15 L   Red-Yellow Podzolic soils, with laterite gravel at or near the surface on old alluvium; undulating.
16   Red-Yellow Podzolic soils on residuum and colluvium from acid rocks; undulating to steep.
P16P   Red-Yellow Podzolic soils on residuum and colluvium from acid rock, on plateaus; gently rolling.

VIII   WELL DRAINED CLAYEY AND LOAMY SOILS, HIGH IN ALUMINUM AND/OR IRON OXIDES, LOW IN BASES, MODERATE TO LOW FERTILITY

17   Reddish-Brown Lateritic soils, on residuum and colluvium from intermediate and basic rocks; undulating to rolling.
     Reddish-Brown Latosols, on residuum and colluvium from basalt; undulating to rolling.
     Red-Yellow Latosols, on old alluvium; undulating to gently rolling.

IX   MISCELLANEOUS SOILS AND LAND TYPES ON HILLS, MOUNTAINS AND PLATEAUS

20   Steep land, intermediate to basic rocks, mainly Red-Yellow Podzolic soils and Reddish Brown Lateritic soils, shallow to deep.
21   Steep land, acid to intermediate rocks, mainly shallow Red-Yellow Podzolic soils
22   Steep land, mainly limestone crags, some Red-Brown Earths.
     Lava plateaus and volcanos, mainly shallow, undifferentiated soils; undulating to steep.

Laboratory analyses

PURPOSE This exercise or series of exercises illustrates the importance of gathering additional data from soil samples collected in the field to quantify further the behavior and properties of soils. Each instructor can decide which analyses are to be emphasized to the students; most courses on soil survey interpretations would probably not include laboratory exercises on all of these techniques outlined here. Although soil profile descriptions and soil maps constitute the most important descriptive information about soils (in the landscape context), laboratory analyses are also of extreme importance for classification and use of the soils. Study Chapter 3 (pages 41–52) of the textbook in preparation for this exercise. Although many separate courses are taught on laboratory analyses of soils, the purpose of this exercise is to emphasize the landscape context of the soil samples and the resultant laboratory data.

SAMPLING Laboratory data are only as precise and reliable as the individual soil samples. Thus, soil samples must be collected very carefully, using clean tools, in accord with soil profile descriptions and soil maps. From the pits dug in the first exercise, collect soil samples of about 1 kilogram from each horizon. Usually, a soil sample is collected from the central portion of each horizon, but samples can be collected at increments of several centimeters if greater detail of soil processes and genesis is desired from the data. Soil samples should be air dried (by spreading on paper towels on tables or shelves) and then thoroughly mixed. They can be stored in plastic bags or glass jars.

Soil samples are collected differently for different purposes. Fertility samples, for example, are commonly collected by taking numerous cores with a soil probe from the plow layer (Ap) of parts of a field within the same soil map unit. Soil cores are placed in a bucket, mixed, and a "composite" sample is analyzed in the laboratory. Results give an "average" analysis for that part of the field. Composite samples should never be collected from several different soil map units of the same field, because those results would be largely meaningless when effects of different texture, fertility, drainage, erosion, slope, and so on, were all mixed

together. Topsoil samples collected from specific points in landscapes are more precise, and have more meaning when several separate analyses are statistically considered; then the effects of soil variability on crop response can be better predicted. Some results of these kinds of analyses are given in Table 14 (page 43) of the textbook. Fertility data from subsoil samples are also valuable to give information to predict performance of deep-rooted crops and plants.

PROCEDURE Collect several separate soil samples from the plow layer (Ap) within different soil map units in the same field. Test each sample for pH using indicators listed on page 38 of the textbook, or the Cornell pH test kit pictured in Figure 8. Usually, this pH test is quite reliable, and relatively few samples are required (see pages 150 and 151 of the textbook). Write down the pH value for each determination, and calculate the range and average value. Considerable differences should be observed between eroded and depositional areas, knobs and seep spots, sandy and clayey places, and so on. Table 8 illustrates computer tabulations for fertility topsoil samples collected from different soils within fields in New York State. Although the range in pH, exchangeable acidity, phosphorus, potassium, and magnesium is great for each soil, there is a characteristic pattern for the data for each soil. Adams, for example, is a sandy acid infertile soil in glacial outwash and most samples are below pH 6.0. Barbour, in contrast, is a fertile silt loam alluvial soil that in most samples has a pH above 6.0.

Considerable emphasis can, of course, be placed on laboratory analyses. Many states and commercial companies have laboratories that analyze soil samples very quickly and give recommendations to farmers and growers. Generally, the laboratories use a weak extract, such as sodium acetate–acetic acid solution, that is leached through the soil sample (Greweling and Peech, 1965), and the "available" nutrients determined are correlated with crop responses for fertilizer recommendations for different soils (Lathwell and Peech, 1973). Numerous soil test kits are also available (Figure 9). The kit pictured in Figure 9 can be used to analyze a variety of different soil samples for soluble salts and cations. From the samples collected from the pits, follow the procedures given in the instruction manual for the kit (Hach, 1973). Samples should be highly contrasting to give students the opportunity to measure a range of different constituents in the soils (e.g., sand and clay, acid and calcareous, low and high organic matter, eroded and depositional areas, etc.)

ENGINEERING SOIL TESTS   The liquid limit device (Figure 10) can be used to illustrate physical soils tests that are important in predicting the engineering behavior of soils. Pages 46–48 of the textbook give the procedure for determining liquid limit, plastic limit, and plasticity index of soil samples (Asphalt Institute, 1969; PCA, 1973). Highly contrasting soil samples should also be selected for these analyses, to emphasize soil sample differences to the students. Thus, clayey and sandy samples will give considerable differences in the plasticity behavior. Differences in organic matter content and clay minerals will influence the test considerably. In conjunction with the tests, students should be given practice in placing their soil samples in the Unified System, and in predicting compressibility, compaction characteristics, and permeability of compacted soil (see page 48 of the textbook).

CHARACTERIZATION ANALYSES   Soil classification (characterization) laboratory analyses are outlined in pages 49–52 of the textbook, and listed in Table 9. These analyses (Soil Survey Staff, 1967) are the most detailed, most expensive, and most complete—and are necessary for the characterization and classification of soils. Analyses listed in Table 9 are designed to show minute differences between horizons and profiles due to pedological processes of soil genesis and weathering. Particle size analyses (Figure 11), for example, are done by removal of organic matter, dispersing the sample, sieving, and pipetting several aliquots as the sediment settles. Each aliquot (sample) is then dried and weighed, and the relative proportions of sand, silt, and clay are calculated. Determination of particles sizes of soil constitutents is one of the most important of all the analyses for prediction of the behavior of a soil.

Throughout the student activities in learning about laboratory analyses, considerable emphasis should be given to considerations about the landscape context of the soils (and samples) and soil variability. It is important to emphasize, for example, that the particle size analyses (Figure 11) are most useful when correlated to determinations made with the fingers in the field by soil scientists making soil profile descriptions and soil maps. A practical exercise such as that illustrated on pages 24–26 of the textbook will help to show students the immediate application of laboratory analyses.

REFERENCES

Asphalt Institute. 1969. Soils manual for design of asphalt pavement structures. Manual Series 10, The Asphalt Institute, College Park, MD. 267 pages.

Greweling, T. and M. Peech. 1965 (revised). Chemical soil tests. Bulletin 960, Agricultural Experiment Station, New York State College of Agriculture and Life Sciences, Cornell University, Ithaca, NY. 59 pages.

Hach. 1973. Instruction manual for Hach soil analysis laboratory Model SA-1. Hach Chemical Company, Ames, IA. 32 pages.

Klausner, S. D. and W. S. Reid. 1979. A summary of soil test results for field crops grown in New York during 1977–1978. Agronomy Mimeo 79-21, Cornell University, Ithaca, NY. 15 pages.

Lathwell, D. J. and M. Peech. 1973 (reprinted). Interpretation of chemical soil tests. Bulletin 995, Agricultural Experiment Station, New York State College of Agriculture and Life Sciences, Cornell University, Ithaca, NY. 40 pages.

PCA. 1973. PCA soil primer. Engineering Bulletin, Portland Cement Association, Skokie, IL. 40 pages.

Soil Survey Staff. 1967. Soil survey laboratory methods and procedures for collecting soil samples. Soil Survey Investigations Report 1, Soil Conservation Service, U.S. Dept. of Agriculture, U.S. Government Printing Office, Washington, DC. 50 pages.

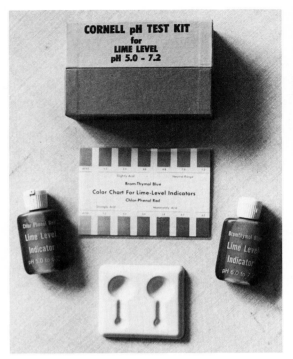

FIGURE 8/*Cornell pH test kit with chlorophenol red and bromthymol blue indicators, useful for soils within the pH range 5.0 to 7.2.*

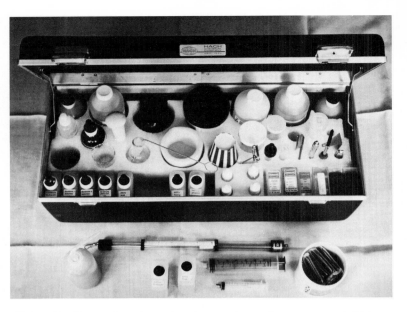

FIGURE 9/*Portable soil test kit that can be easily transported. This kit can be used to analyze soils for a wide range of soluble salts and cations (see pages 50–52 of the textbook).*

FIGURE 10/*Liquid limit device for determining Atterberg limits and the plasticity index of a soil, for classification in the Unified System (see pages 46–48 of the textbook).*

FIGURE 11/*Withdrawal of soil sediment samples with a pipette during particle size analyses. The dispersed sediment settles differentially, so that soil fractions of different weights and sizes can be separated by aliquot withdrawal at different time intervals (photo by Ken Olson).*

TABLE 8/Soil test results tabulated by computer and arranged according to high, medium, and low values (adapted from Klausner and Reid, 1979).

Soil Series Name	pH AVE	pH <5.6 %	pH 5.6–6.0 %	pH 6.0–6.6 %	pH >6.6 %	EXC. ACIDITY AVE (me/100g)	<10 %	10–20 %	>20 %	PHOSPHORUS AVE (lb/A)	P VL %	P LO %	P MD %	P HI %	P VH %	POTASSIUM AVE (lb/A)	K VL %	K LO %	K MD %	K HI %	K VH %	MAGNESIUM AVE (lb/A)	Mg VL %	Mg LO %	Mg MD %	Mg HI %	Mg VH %	CALCIUM AVE (lb/A)
ADAMS	5.9	30	25	30	15	17	0	82	18	5	0	69	21	5	5	90	25	45	30	0	0	124	15	30	20	15	20	2520
ALTON	6.1	16	32	33	19	14	17	80	3	12	0	28	28	40	4	206	3	20	23	23	31	138	7	16	15	42	20	2608
AMENIA	6.5	2	20	40	38	13	8	92	0	6	0	58	24	18	0	135	4	43	25	13	15	145	0	5	24	58	13	4963
ANGOLA	6.5	0	13	47	40	9	50	50	0	4	0	66	27	0	0	128	1	13	0	74	13	348	0	0	0	13	87	5400
APPLETON	6.3	9	30	33	28	14	17	66	17	7	0	58	22	17	3	161	1	10	16	39	34	298	0	0	6	12	78	3918
ARKPORT	6.6	6	14	16	64	11	43	57	0	18	0	14	19	54	13	204	0	9	33	21	34	253	1	4	6	17	65	2647
ARNOT	6.0	26	21	42	11	16	0	89	11	5	0	57	11	32	0	133	11	26	9	55	16	220	5	16	0	32	47	2077
AURORA	5.9	18	36	37	9	11	33	67	0	3	0	91	0	9	0	122	0	18	9	55	18	290	0	0	0	50	73	2133
BARBOUR	6.3	13	17	41	29	10	57	43	0	34	0	4	29	38	29	199	0	13	29	25	33	169	0	0	21	24	29	3063
BATH	6.1	18	26	36	20	15	8	83	9	13	0	35	27	31	7	191	3	7	23	28	39	263	0	5	9	26	62	2709
BERNARDSTON	6.1	19	24	40	17	12	17	81	2	13	0	27	28	37	8	189	3	20	21	26	30	251	0	5	5	54	64	2277
BOMBAY	6.1	33	13	34	20	13	43	57	0	9	0	33	40	20	7	157	0	27	33	20	38	177	0	13	5	23	33	2978
BRACEVILLE	6.2	0	38	47	15	13	1	99	0	10	0	38	46	8	8	202	0	8	46	8	14	286	0	0	0	18	77	2458
BROADALBIN	6.1	21	11	54	14	20	0	33	67	7	0	79	10	6	5	115	32	21	29	4	25	232	0	32	0	28	50	1852
BURDETT	6.1	17	30	41	12	15	5	79	16	5	0	76	11	11	2	142	1	13	23	41	25	314	0	20	3	28	66	3447
CAMRODEN	5.7	42	41	13	4	19	0	56	44	4	0	68	17	15	0	199	4	13	17	41	43	162	0	0	13	43	24	2647
CANANDAIGUA	6.4	5	48	14	33	12	9	91	0	7	0	33	38	29	0	105	5	19	43	33	0	345	0	0	5	10	85	4217
CASTILE	5.4	50	25	25	0	13	38	49	13	10	0	50	8	42	0	203	8	34	8	17	33	197	0	20	0	6	25	1706
CAZENOVIA	6.5	5	19	32	44	9	57	43	0	11	0	36	30	27	7	156	0	0	18	57	31	348	0	0	2	14	92	3655
CHARLTON	6.1	11	36	42	11	14	1	99	0	7	0	32	32	42	2	155	11	28	18	25	18	324	0	0	1	36	75	1871
CHENANGO	5.8	30	26	32	12	16	6	76	18	11	0	26	30	36	0	221	0	9	11	30	50	223	0	4	7	42	50	2738
CHIPPEWA	6.1	8	25	59	8	15	1	99	0	11	0	67	25	5	0	133	0	33	25	34	8	270	0	5	9	35	58	2791
CHURCHVILLE	6.5	24	12	29	35	12	1	99	0	6	0	47	41	12	0	100	0	24	24	40	12	338	0	0	0	27	82	5025
COLLAMER	6.3	5	22	50	23	11	16	84	0	8	0	31	44	24	1	156	0	9	27	45	20	234	0	0	12	27	50	3870
COLONIE	6.1	14	32	45	9	11	80	10	10	12	0	14	36	50	0	234	14	36	32	0	18	109	5	3	12	22	14	935
COLOSSE	6.0	37	32	9	27	16	0	86	14	6	0	55	27	18	0	156	12	18	18	28	27	131	0	31	23	10	27	2230
COLTON	6.2	10	27	51	12	13	10	80	10	7	0	42	31	27	0	172	12	12	27	19	30	124	0	18	28	1	23	2150
CONESUS	6.3	11	20	43	26	13	3	90	7	8	0	32	40	25	3	144	0	4	42	42	20	294	0	35	15	54	74	3610
COPAKE	6.5	5	10	50	35	11	33	67	0	9	0	37	37	26	0	120	35	30	10	5	20	333	0	1	5	24	85	2070
COVINGTON	6.3	10	10	70	10	21	0	71	29	5	0	99	1	0	0	116	0	10	30	30	30	386	0	0	7	29	99	4540
DALTON	5.9	31	15	54	0	17	0	83	17	3	0	69	23	8	0	132	0	31	30	30	18	241	8	0	0	21	46	2971
DARIEN	6.3	10	14	48	28	14	13	81	6	8	0	62	17	8	0	126	0	0	31	53	8	316	0	0	5	8	71	4230
DUNKIRK	6.4	19	14	24	43	10	57	43	0	10	0	60	25	17	4	170	0	5	38	24	33	304	0	1	4	31	71	2907
ELMWOOD	6.1	14	36	29	21	11	29	77	15	6	0	43	36	5	10	130	14	29	21	21	7	277	0	0	7	21	72	2342
EMPEYVILLE	5.8	25	46	29	0	18	14	71	29	3	0	67	33	0	0	92	17	42	29	8	7	134	8	29	0	38	25	2025
ERIE	6.1	14	38	30	18	14	14	78	8	11	0	35	28	31	6	167	1	11	37	37	26	275	0	0	5	30	74	2590
FARMINGTON	6.3	10	19	46	25	15	13	93	7	11	0	54	17	19	10	154	10	15	23	23	29	296	0	1	18	12	67	4781
FREMONT	6.1	17	28	34	21	14	8	77	15	6	0	59	24	10	7	163	0	24	14	23	38	296	0	4	2	31	69	2679
GALEN	6.4	6	32	31	31	10	50	50	0	9	0	38	24	38	0	193	0	38	14	25	31	232	0	10	13	21	49	3423
GLOUCESTER	6.0	30	30	30	10	18	0	67	33	4	0	50	40	10	0	160	30	10	10	10	38	124	0	30	30	38	78	3514
HAMLIN	6.4	7	26	41	26	11	37	63	0	12	0	19	35	39	7	175	10	5	29	30	51	290	0	20	30	30	66	4071
HERKIMER	6.0	22	27	40	11	15	3	89	8	8	0	27	49	22	2	218	0	14	31	31	50	242	0	7	3	20	75	4294
HILTON	6.3	11	22	36	31	12	24	76	0	9	0	30	38	30	2	180	10	20	32	38	38	290	0	3	5	17	75	3396

TABLE 9/*Laboratory methods for characterization of samples collected from soil profile horizons, and for classification of soils in Soil Taxonomy (Soil Survey Staff, 1967).*

1. SAMPLE COLLECTION AND PREPARATION
   A. Field sampling
      1. Site selection
      2. Soil sampling
         a. Stony soils
   B. Laboratory preparation
      1. Standard (airdry)
         a. Square-hole 2-mm sieve
         b. Round-hole 2-mm sieve
      2. Field moist
      3. Carbonate-containing material
      4. Carbonate-indurated material
2. CONVENTIONS
   A. Size-fraction base for reporting
      1. <2-mm
      2. <size specified
   B. Data-sheet symbols
      tr: trace, not measurable by quantitative procedure used or less than reportable amount
      tr(s): trace, detectable only by qualitative procedure more sensitive than quantitative procedure used
      -- : analysis run but none detected
      -(s): none detected by sensitive qualitative test
      blank: analysis not run
      nd: analysis not run
      <: less than reported amount or none present
3. PARTICLE-SIZE ANALYSES
   A. <2-mm fraction (pipet method)
      1. Airdry samples
         a. Carbonate and noncarbonate clay
      2. Moist samples
         a. Carbonate and noncarbonate clay
   B. >2-mm fraction
      1. Weight estimates
      2. Volume estimates
4. FABRIC-RELATED ANALYSES
   A. Bulk density
      1. Saran-coated clods
         a. Field state
         b. Airdry
         c. 30-cm absorption
         d. 1/3-bar desorption I
         e. 1/3-bar desorption II
         f. 1/3-bar desorption III
         g. 1/10-bar desorption
         h. Ovendry
      2. Paraffin-coated clods
         a. Ovendry
      3. Cores
         a. Field moist
      4. Nonpolar-liquid-saturated clods
   B. Water retention
      1. Pressure-plate extraction (1/3 or 1/10 bar)
         a. Sieved samples
         b. Soil pieces
         c. Natural clods
         d. Cores
      2. Pressure-membrane extraction (15 bars)
      3. Sand table absorption
      4. Field state
      5. Airdry
   C. Water-retention difference
      1. 1/3 bar to 15 bars
      2. 1/10 bar to 15 bars
   D. Coefficient of linear extensibility
      1. Dry to moist
   E. Micromorphology
      1. Thin sections
         a. Preparation
         b. Interpretation
         c. Moved-clay percentage
5. ION-EXCHANGE PROPERTIES
   A. Cation-exchange capacity
      1. NH_4OAc, pH 7.0
         a. Direct distillation
         b. Displacement, distillation

5A. Cation-exchange capacity (cont.)
      2. NaOAc, pH 8.2
         a. Centrifuge method
      3. Sum of cations
         a. Acidity by $BaCl_2$-TEA, pH 8.2; bases by NH_4OAc, pH 7.0
      4. KOAc, pH 7.0
      5. $BaCl_2$, pH 8.2
         a. Barium by flame photometry
   B. Extractable bases
      1. NH_4OAc extraction
         a. Uncorrected
         b. Corrected (exchangeable)
      2. KCl-TEA extraction, pH 8.2
   C. Base saturation
      1. NH_4OAc, pH 7.0
      2. NaOAc, pH 8.2
      3. Sum of cations
   D. Sodium saturation (exchangeable Na pct.)
      1. NaOAc, pH 8.2
      2. NH_4OAc, pH 7.0
   E. Sodium adsorption ratio
6. CHEMICAL ANALYSES
   A. Organic carbon
      1. Acid dichromate digestion
         a. $FeSO_4$ titration
         b. CO_2 evolution, gravimetric
      2. Dry combustion
         a. CO_2 evolution I
         b. CO_2 evolution II
      3. Peroxide digestion
         a. Weight loss
   B. Nitrogen
      1. Kjeldahl digestion
         a. Ammonia distillation
      2. Semimicro Kjeldahl
         a. Ammonia distillation
   C. Iron
      1. Dithionite extraction
         a. Dichromate titration
         b. EDTA titration
      2. Dithionite-citrate extraction
         a. Orthophenanthroline colorimetry
      3. Dithionite-citrate-bicarbonate extraction
         a. Potassium-thiocyanate colorimetry
      4. Pyrophosphate-dithionite extraction
   D. Manganese
      1. Dithionite extraction
         a. Permanganate colorimetry
   E. Calcium carbonate
      1. HCl treatment
         a. Gas volumetric
         b. Manometric
         c. Weight loss
         d. Weight gain
         e. Titrimetric
      2. Sensitive qualitative method
         a. Visual, gas bubbles
   F. Gypsum
      1. Water extract
         a. Precipitation in acetone
   G. Aluminum
      1. KCl extraction I, 30 min
         a. Aluminon I
         b. Aluminon II
         c. Aluminon III
         d. Fluoride titration
      2. KCl extraction II, overnight
         a. Aluminon I
      3. NH_4OAc extraction
         a. Aluminon III
      4. NaOAc extraction
         a. Aluminon III
   H. Extractable acidity
      1. $BaCl_2$-triethanolamine I
         a. Back-titration with HCl
      2. $BaCl_2$-triethanolamine II
         a. Back-titration with HCl
      3. KCl-triethanolamine
         a. Back-titration with NaOH
   I. Carbonate
      1. Saturation extract
         a. Acid titration

6. CHEMICAL ANALYSES (cont.)
   J. Bicarbonate
      1. Saturation extract
         a. Acid titration
   K. Chloride
      1. Saturation extract
         a. Mohr titration
         b. Potentiometric titration
   L. Sulfate
      1. Saturation extract
         a. Gravimetric, $BaSO_4$
      2. NH_4OAc extraction
         a. Gravimetric, $BaSO_4$
   M. Nitrate
      1. Saturation extract
         a. PDS acid colorimetry
   N. Calcium
      1. Saturation extract
         a. EDTA titration
      2. NH_4OAc extraction
         a. EDTA-alcohol separation
         b. Oxalate-permanganate I
         c. Oxalate-permanganate II Fe, Al, and Mn removed
         d. Oxalate-cerate
      3. NH_4Cl-EtOH extraction
         a. EDTA titration
      4. KCl-TEA extraction
         a. Oxalate-permanganate
   O. Magnesium
      1. Saturation extract
         a. EDTA titration
      2. NH_4OAc extraction
         a. EDTA-alcohol separation
         b. Phosphate titration
         c. Gravimetric, $Mg_2P_2O_7$
      3. NH_4Cl-EtOH extraction
         a. EDTA titration
   P. Sodium
      1. Saturation extract
         a. Flame photometry
      2. NH_4OAc extraction
         a. Flame photometry
   Q. Potassium
      1. Saturation extract
         a. Flame photometry
      2. NH_4OAc extraction
         a. Flame photometry
   R. Sulfur
      1. $NaHCO_3$ extraction, pH 8.5
         a. Methylene blue
   S. Total phosphorus
      1. Perchloric-acid digestion
         a. Molybdovanadophosphoric-acid colorimetry
7. MINERALOGY
   A. Instrumental analysis
      1. Preparation
         a. Carbonate removal
         b. Organic-matter removal
         c. Iron removal
         d. Particle-size fractionation
      2. X-ray diffraction
      3. Differential thermal analysis
   B. Optical analysis
      1. Grain studies
   C. Total analysis
      1. Chemical
      2. X-ray emission spectrography
   D. Surface area
      1. Glycerol retention
8. MISCELLANEOUS
   A. Saturated paste, mixed
      1. Saturation extract
         a. Conductivity
      2. Conductivity, saturated paste
   B. Saturated paste, capillary rise
      1. Saturation extract
         a. Conductivity
   C. pH
      1. Soil suspensions
         a. Water dilution
         b. Saturated paste
         c. KCl
   D. Ratios
      1. To total clay
      2. To noncarbonate clay
      3. Ca to Mg (extractable)

Soil Taxonomy

PURPOSE This exercise is intended to acquaint the student with some of the details and generalizations of Soil Taxonomy (Soil Survey Staff, 1975, 1980). References cited in this section should be studied carefully. No hesitation should be involved in seeking help from local and state soil scientists, and in naming and classifying soils already described, mapped, and analyzed in the previous exercises. Procedures for better understanding and using our soils involve first detailed description, analyses, and delineation and then generalization (classification) to make the information relevant to understanding and solving soils problems. Pages 53–62 of the textbook provide an introduction to the nomenclature and some of the principles of Soil Taxonomy (Soil Survey Staff, 1975).

PROCEDURE Investigate all the available information relevant to classification of the soils in your area. Published soil survey reports generally contain a section that lists the soils and the Soil Taxonomy classification. State soil survey offices and experiment stations have copies of the master list of all the soil series and Soil Taxonomy classification of each state in the United States and for Puerto Rico and the Virgin Islands (Soil Survey Staff, 1980); other countries have similar lists available. Table 10 provides an illustration of one page from the master list of more than 11,000 soil series (Soil Survey Staff, 1980), including the state most closely associated with each soil and the Soil Taxonomy Family name. All the information is constantly correlated as new data become available; thus, soil maps and legend names (illustrated in Figs. 2 and 3 and Table 6) can readily be related to the official classification. All that is needed for this exercise is a soil map with current soil names and the master list, together with the supportive soil profile descriptions and laboratory data.

Buol et al. (1980) provide good discussions of soil genesis and classification, have separate chapters on the Orders of Soil Taxonomy, and relate Soil Taxonomy to some of the other prominent soil classification systems used in various countries. Although each soil classification system has its merits, Soil Taxonomy is the most detailed and quantitative, and is widely used in many countries. The soil classification system used by the Food and Agriculture Organization of the United Nations is essentially the Soil Taxonomy system with some modifications to achieve a unified legend for the World Soil Map (see pages 143–146 of the textbook). At present, considerable activity is under way to revise and improve Soil Taxonomy, especially in tropical areas in developing countries.

RESEARCH Excellent illustrations are available to demonstrate the importance of complete and detailed information for classification of the soils. The voluminous "Desert Project Soil Monograph" (Gile and Grossman, 1979) is a compilation of 15 years of research into the geomorphology and pedology of the area around Las Cruces, New Mexico. Figure 12 is a diagram of the study area. Table 11, for example, is a soil profile description for the Vinton soil from the north bank of an arroyo (gully) in Organ alluvium derived from monzonite, classified Typic Torrifluvent; sandy, mixed, thermic. Table 12 illustrates the format for reporting laboratory data, according to the procedures listed in Table 9. The monograph also contains many data on other soils of the area, and much information about the climate and geomorphology of the area.

CORRELATIONS Efforts under way to improve and expand Soil Taxonomy are illustrated in the Proceedings of the First International Soil Classification Workshop (Camargo and Beinroth, 1978). The workshop was held in Brazil and involved soils experts from many countries. Participants in the workshop examined the adequacy of Soil Taxonomy with respect to tropical soils, proposed pertinent changes in Soil Taxonomy, identified relevant knowledge gaps and research needs, finalized new definitions for certain categories of Alfisols and Ultisols, and studied critical examples of the soils in the field. An example of one of the soils studied is given in Table 13, and Table 14 gives laboratory data for soil horizon samples taken from that soil. Table 13 has classifications given for the soil from Soil Taxonomy, the FAO System, and the French Soil Classification System.

Complete descriptions and data, of course, are not always available for all soils. Thus, it is advisable to give students experiences in classifying soils with limited data, where assumptions need to be made. Typically, soils are classified with limited data, and revisions in the classifications are made as more data become available and as research projects are completed. Student classification of soils in their project area will give them some experience in the categorization of soils according to their physical, chemical, and climatic properties.

CLASSIFICATION BY COMPUTER   If computers are available, students should experiment with classification exercises according to programs written by Robert and Rust (1981). Use of a microcomputer can greatly ease and improve speed and accuracy of soil classification according to Soil Taxonomy. This first development (Robert and Rust, 1981) tests soil data for diagnostic horizon identification. In a first method, selection options are displayed sequentially on the computer video screen. Yes/No decisions are made depending on the matching of the morphologic, physical, and chemical properties of the soils. Responses determine the succeeding steps of the "selection tree" until the horizon is accepted or rejected as a particular diagnostic horizon. In a second method, all the properties, such as soil, climate, and mineralogy, required for decisions for classification are first displayed on the computer screen. When all the needed data are available, they can be interactively entered into the computer. Then an internal test is computed and the horizon is accepted or rejected as a member of the classification category in question. If rejected, the nondefining soil property is displayed by the computer (Robert and Rust, 1981).

GENERAL SOIL MAPS   Examples of generalizations of soil information according to Soil Taxonomy are given in Figures 13, 14, and 4.

Figure 13 is a general soil map of the world showing Orders and Suborders according to Soil Taxonomy. Figure 14 is a general soil map of the United States showing Orders and Suborders, with a legend that has more detail of explanation of the map units. Figure 4 is a general soil map (soil association map) at a larger scale, and Great Groups are designated in the Soil Taxonomy legend nomenclature. Soil Taxonomy permits categorization of a natural classification of soils to be made conveniently at different levels of map scale and different degrees of detail and generalization. These different levels are extremely useful for teaching and learning about soils.

REFERENCES

Buol, S. W., F. D. Hole, and R. J. McCracken. 1980 (second edition). Soil genesis and classification. Iowa State University Press, Ames, IA. 404 pages.

Camargo, M. N. and F. H. Beinroth (Editors). 1978. First international soil classification workshop: Proceedings of a workshop held in Rio de Janeiro, Brazil, from June 20 to July 1, 1977 with soil study tours in the states of Rio de Janeiro, Parana, Sergipe, Alagoas, and Pernambuco. Servico Nacional de Levantamento e Conservacao de Solos, Rio de Janeiro, Brazil. 376 pages.

Gile, L. H. and R. B. Grossman. 1979. The desert project soil monograph: Soils and landscapes of a desert region astride the Rio Grande Valley near Las Cruces, New Mexico. Soil Conservation Service, U.S. Dept. of Agriculture, Washington, DC. 984 pages.

Robert, P. C. and R. H. Rust. 1981. Development of a microcomputer interactive access to Soil Taxonomy (A. Diagnostic Horizon). Agronomy Abstracts 1981: 205.

Soil Survey Staff. 1975. Soil Taxonomy: A basic system of soil classification for making and interpreting soil surveys. Agriculture Handbook 436, U.S. Dept. of Agriculture, U.S. Government Printing Office, Washington, DC. 754 pages.

Soil Survey Staff. 1980. Classification of soil series of the United States, Puerto Rico, and the Virgin Islands. Soil Conservation Service, U.S. Dept. of Agriculture, Washington, DC. 330 pages.

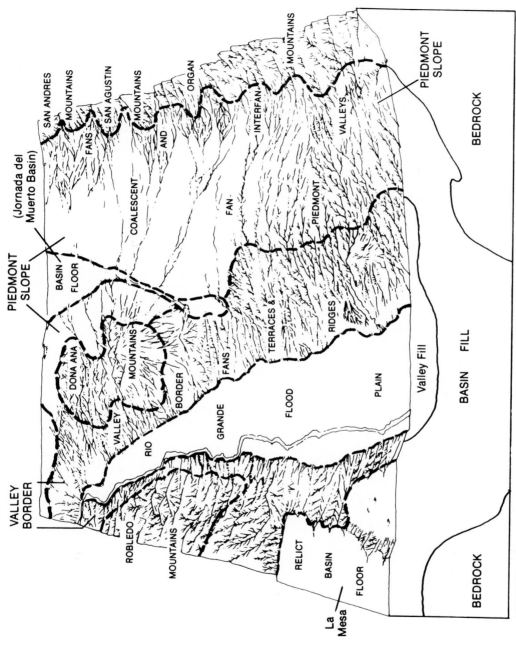

FIGURE 12/*Diagram of the landforms around Las Cruces, New Mexico, studied in the Desert Soil-Geomorphology Project. The Rio Grande floodplain is about 5 miles (8 km) wide at the cross section (adapted from Gile and Grossman, 1979).*

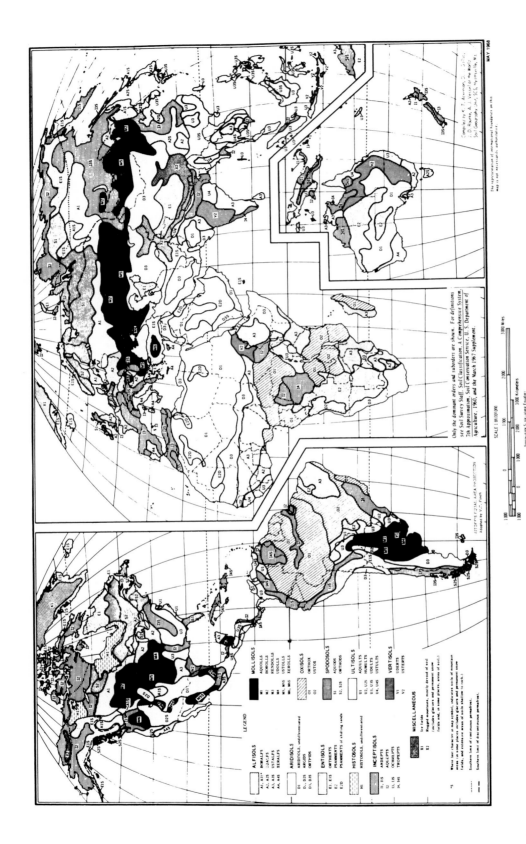

FIGURE 13/*General soil map of the world, showing Orders and Suborders according to Soil Taxonomy (map from Soil Survey Division, Soil Conservation Service, U.S. Department of Agriculture).*

FIGURE 14/*General soil map of the United States, showing Orders and Suborders according to Soil Taxonomy (map from Soil Survey Division, Soil Conservation Service, U.S. Department of Agriculture).*

LEGEND

Only the dominant orders and suborders are shown. Each delineation has many inclusions of other kinds of soil. General definitions for the orders and suborders follow. For complete definitions see Soil Survey Staff, Soil Classification, A Comprehensive System, 7th Approximation, Soil Conservation Service, U. S. Department of Agriculture, 1960 (for sale by U. S. Government Printing Office) and the March 1967 supplement (available from Soil Conservation Service, U. S. Department of Agriculture). Approximate equivalents in the modified 1938 soil classification system are indicated for each suborder.

 **ALFISOLS . . .** Soils with gray to brown surface horizons, medium to high base supply, and subsurface horizons of clay accumulation; usually moist but may be dry during warm season

A1   AQUALFS (seasonally saturated with water) gently sloping; general crops if drained, pasture and woodland if undrained (Some Low–Humic Gley soils and Planosols)

A2   BORALFS (cool or cold) gently sloping; mostly woodland, pasture, and some small grain (Gray Wooded soils)

A2S BORALFS steep; mostly woodland

A3   UDALFS (temperate or warm, and moist) gently or moderately sloping; mostly farmed, corn, soybeans, small grain, and pasture (Gray–Brown Podzolic soils)

A4   USTALFS (warm and intermittently dry for long periods) gently or moderately sloping; range, small grain, and irrigated crops (Some Reddish Chestnut and Red–Yellow Podzolic soils)

A5S XERALFS (warm and continuously dry in summer for long periods, moist in winter) gently sloping to steep; mostly range, small grain, and irrigated crops (Noncalcic Brown soils)

 ARIDISOLS . . . Soils with pedogenic horizons, low in organic matter, and dry more than 6 months of the year in all horizons

D1   ARGIDS (with horizon of clay accumulation) gently or moderately sloping; mostly range, some irrigated crops (Some Desert, Reddish Desert, Reddish–Brown, and Brown soils and associated Solonetz soils)

D1S ARGIDS gently sloping to steep

D2   ORTHIDS (without horizon of clay accumulation) gently or moderately sloping; mostly range and some irrigated crops (Some Desert, Reddish Desert, Sierozem, and Brown soils, and some Calcisols and Solonchak soils)

D2S ORTHIDS gently sloping to steep

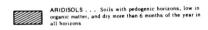 **ENTISOLS . . .** Soils without pedogenic horizons

E1   AQUENTS (seasonally saturated with water) gently sloping; some grazing

E2   ORTHENTS (loamy or clayey textures) deep to hard rock; gently to moderately sloping; range or irrigated farming (Regosols)

E3   ORTHENTS shallow to hard rock; gently to moderately sloping; mostly range (Lithosols)

E3S ORTHENTS shallow to hard rock; steep; mostly range

E4   PSAMMENTS (sand or loamy sand textures) gently to moderately sloping; mostly range in dry climates, woodland or cropland in humid climates (Regosols)

FIGURE 14 *(continued)*

 HISTOSOLS . . . Organic soils

H1   FIBRISTS (fibrous or woody peats, largely undecomposed) mostly wooded or idle (Peats)

H2   SAPRISTS (decomposed mucks) truck crops if drained, idle if undrained (Mucks)

 INCEPTISOLS . . . Soils that are usually moist, with pedogenic horizons of alteration of parent materials but not of accumulation

I1S ANDEPTS (with amorphous clay or vitric volcanic ash and pumice) gently sloping to steep; mostly woodland; in Hawaii mostly sugar cane, pineapple, and range (Ando soils, some Tundra soils)

I2   AQUEPTS (seasonally saturated with water) gently sloping; if drained, mostly row crops, corn, soybeans, and cotton; if undrained, mostly woodland or pasture (Some Low–Humic Gley soils and Alluvial soils)

I2P AQUEPTS (with continuous or sporadic permafrost) gently sloping to steep; woodland or idle (Tundra soils)

I3   OCHREPTS (with thin or light–colored surface horizons and little organic matter) gently to moderately sloping; mostly pasture, small grain, and hay (Sols Bruns Acides and some Alluvial soils)

I3S OCHREPTS gently sloping to steep; woodland, pasture, small grains

I4S UMBREPTS (with thick dark–colored surface horizons rich in organic matter) moderately sloping to steep; mostly woodland (Some Regosols)

 MOLLISOLS . . . Soils with nearly black, organic–rich surface horizons and high base supply

M1   AQUOLLS (seasonally saturated with water) gently sloping; mostly drained and farmed (Humic Gley soils)

M2   BOROLLS (cool or cold) gently or moderately sloping, some steep slopes in Utah; mostly small grain in North Central States, range and woodland in Western States (Some Chernozems)

M3   UDOLLS (temperate or warm, and moist) gently or moderately sloping; mostly corn, soybeans, and small grains (Some Brunizems)

M4   USTOLLS (intermittently dry for long periods during summer) gently to moderately sloping; mostly wheat and range in western part, wheat and corn or sorghum in eastern part, some irrigated crops (Chestnut soils and some Chernozems and Brown soils)

M4S USTOLLS moderately sloping to steep; mostly range or woodland

M5   XEROLLS (continuously dry in summer for long periods, moist in winter) gently to moderately sloping; mostly wheat, range, and irrigated crops (Some Brunizems, Chestnut, and Brown soils)

M5S XEROLLS moderately sloping to steep; mostly range

SPODOSOLS . . . Soils with accumulations of amorphous materials in subsurface horizons

S1   AQUODS (seasonally saturated with water) gently sloping; mostly range or woodland; where drained in Florida, citrus and special crops (Ground–Water Podzols)

S2   ORTHODS (with subsurface accumulations of iron, aluminum, and organic matter) gently to moderately sloping; woodland, pasture, small grains, special crops (Podzols, Brown Podzolic soils)

S2S ORTHODS steep; mostly woodland

 **ULTISOLS . . .** Soils that are usually moist with horizon of clay accumulation and a low base supply

U1   AQUULTS (seasonally saturated with water) gently sloping, woodland and pasture if undrained, feed and truck crops if drained (Some Low–Humic Gley soils)

U2S HUMULTS (with high or very high organic–matter content) moderately sloping to steep; woodland and pasture if steep, sugar cane and pineapple in Hawaii, truck and seed crops in Western States (Some Reddish–Brown Lateritic soils)

U3   UDULTS (with low organic–matter content; temperate or warm, and moist) gently to moderately sloping; woodland, pasture, feed crops, tobacco, and cotton (Red–Yellow Podzolic soils, some Reddish–Brown Lateritic soils)

U3S UDULTS moderately sloping to steep; woodland, pasture

U4S XERULTS (with low to moderate organic–matter content, continuously dry for long periods in summer) range and woodland (Some Reddish–Brown Lateritic soils)

VERTISOLS . . . Soils with high content of swelling clays and wide deep cracks at some season

V1   UDERTS (cracks open for only short periods, less than 3 months in a year) gently sloping; cotton, corn, pasture, and some rice (Some Grumusols)

V2   USTERTS (cracks open and close twice a year and remain open more than 3 months); general crops, range, and some irrigated crops (Some Grumusols)

AREAS with little soil . . .

X1   Salt flats

X2   Rockland, ice fields

NOMENCLATURE

The nomenclature is systematic. Names of soil orders end in *sol* (L. *solum*, soil), e. g., ALFISOL, and contain a formative element used as the final syllable in names of taxa in suborders, great groups, and sub-groups.

Names of suborders consist of two syllables, e. g., AQUALF. Formative elements in the legend for this map and their connotations are as follows

and	– Modified from Ando soils; soils from vitreous parent materials
aqu	– L. *aqua*, water; soils that are wet for long periods
arg	– Modified from L. *argilla*, clay; soils with a horizon of clay accumulation
bor	– Gr. *boreas*, northern; cool
fibr	– L. *fibra*, fiber; least decomposed
hum	– L. *humus*, earth; presence of organic matter
ochr	– Gr. base of ochros, pale; soils with little organic matter
orth	– Gr. *orthos*, true; the common or typical
psamm	– Gr. *psammos*, sand; sandy soils
sapr	– Gr. *sapros*, rotten; most decomposed
ud	– L. *udus*, humid; of humid climates
umbr	– L. *umbra*, shade; dark colors reflecting much organic matter
ust	– L. *ustus*, burnt; of dry climates with summer rains
xer	– Gr. *xeros*, dry; of dry climates with winter rains

AUGUST 19

TABLE 10/*Page from computer master list of soil series and Soil Taxonomy Family classification of soil series of the United States, Puerto Rico, and the Virgin Islands (Adapted from Soil Survey Staff, 1980).*

SERIES	STATE	SOIL FAMILY
ALAELOA	HI	CLAYEY, OXIDIC, ISOHYPERTHERMIC ORTHOXIC TROPOHUMULTS
ALAGA	AL	THERMIC, COATED TYPIC QUARTZIPSAMMENTS
ALAKAI	HI	CLAYEY, KAOLINITIC, DYSIC, ISOMESIC TERRIC TROPOSAPRISTS
ALAHA	NM	FINE-SILTY, MIXED, THERMIC USTOLLIC CAMBORTHIDS
ALAMANCE	SC	FINE-SILTY, SILICEOUS, THERMIC TYPIC HAPLUDULTS
ALAMO	CA	FINE, MONTMORILLONITIC, THERMIC TYPIC DURAQUOLLS
ALAMOGORDO	NM	COARSE-LOAMY, GYPSIC, THERMIC TYPIC GYPSIORTHIDS
ALAMOSA	CO	FINE-LOAMY, MIXED, FRIGID TYPIC ARGIAQUOLLS
ALAPAHA	GA	LOAMY, SILICEOUS, THERMIC ARENIC PLINTHIC PALEAQUULTS
ALAPAI	HI	THIXOTROPIC, ISOTHERMIC TYPIC HYDRANDEPTS
ALAZAN	TX	FINE-LOAMY, SILICEOUS, THERMIC AQUIC GLOSSUDALFS
ALBAN	WI	COARSE-LOAMY, MIXED TYPIC GLOSSOBORALFS
ALBANO	VA	FINE, MIXED, MESIC TYPIC OCHRAQUALFS
ALBANY	GA	LOAMY, SILICEOUS, THERMIC GROSSARENIC PALEUDULTS
ALBATON	IA	FINE, MONTMORILLONITIC (CALCAREOUS), MESIC VERTIC FLUVAQUENTS (VERY-FINE)
ALBEE	OR	FINE-LOAMY, MIXED, FRIGID ULTIC HAPLOXEROLLS
ALBEMARLE	VA	FINE-LOAMY, MIXED, MESIC TYPIC HAPLUDULTS
ALBERTVILLE	AL	CLAYEY, MIXED, THERMIC TYPIC HAPLUDULTS
ALBINAS	WY	FINE-LOAMY, MIXED, MESIC, PACHIC ARGIUSTOLLS
ALBION	KS	COARSE-LOAMY, MIXED, THERMIC UDIC ARGIUSTOLLS
ALBRIGHTS	PA	FINE-LOAMY, MIXED, MESIC AQUIC FRAGIUDALFS
ALCESTER	SD	FINE-SILTY, MIXED, MESIC CUMULIC HAPLUSTOLLS
ALCOA	TN	CLAYEY, OXIDIC, THERMIC RHODIC PALEUDULTS
ALCONA	MI	COARSE-LOAMY, MIXED, FRIGID ALFIC HAPLORTHODS
ALCORN	NV	COARSE-LOAMY, MIXED, MESIC TYPIC CAMBORTHIDS
ALCOT	OR	CINDERY, MESIC TYPIC XERORTHENTS
ALCOVA	WY	FINE-LOAMY, MIXED BOROLLIC HAPLARGIDS
ALDA	NE	COARSE-LOAMY, MIXED, MESIC FLUVAQUENTIC HAPLUSTOLLS
ALDAX	NV	LOAMY-SKELETAL, MIXED, MESIC LITHIC HAPLOXEROLLS
ALDEN	NY	FINE-LOAMY, MIXED, NONACID, MESIC MOLLIC HAPLAQUEPTS
ALDER	MT	FINE, MIXED UDIC ARGIBOROLLS
ALDERDALF	WA	SANDY-SKELETAL, MIXED, MESIC TYPIC TORRIORTHENTS
ALDERWOOD	WA	LOAMY-SKELETAL, MIXED, MESIC DYSTRIC ENTIC DUROCHREPTS
ALDINE	TX	FINE-SILTY OVER CLAYEY, SILICEOUS, THERMIC AERIC GLOSSAQUALFS
ALDING	OR	CLAYEY, MONTMORILLONITIC, FRIGID LITHIC ULTIC ARGIXEROLLS
ALDINO	MD	FINE-SILTY, MIXED, MESIC TYPIC FRAGIUDALFS
ALEDO	TX	LOAMY-SKELETAL, CARBONATIC, THERMIC LITHIC HAPLUSTOLLS
ALEKNAGIK	AK	MEDIAL HUMIC CRYORTHODS
ALEMEDA	NM	LOAMY-SKELETAL, MIXED, THERMIC TYPIC CALCIORTHIDS
ALEX	ID	FINE-LOAMY OVER SANDY OR SANDY-SKELETAL, MIXED PACHIC CRYOBOROLLS
ALEXANDRIA	OH	FINE, ILLITIC, MESIC TYPIC HAPLUDALFS
ALFIR	WA	MEDIAL-SKELETAL DYSTRIC CRYANDEPTS
ALFORD	IN	FINE-SILTY, MIXED, MESIC TYPIC HAPLUDALFS
ALGANSEE	MI	MIXED, MESIC AQUIC UDIPSAMMENTS
ALGARROBO	PR	COARSE-LOAMY, SILICEOUS, ISOHYPERTHERMIC ENTIC TROPOHUMODS
ALGERITA	NM	COARSE-LOAMY, MIXED, THERMIC TYPIC CALCIORTHIDS
ALGIERS	OH	FINE-LOAMY, MIXED, NONACID, MESIC AQUIC UDIFLUVENTS
ALGOMA	OR	MEDIAL OVER SANDY OR SANDY-SKELETAL, MIXED (CALCAREOUS), MESIC MOLLIC HALAQUEPTS
ALHAMBRA	NV	COARSE-LOAMY, MIXED (CALCAREOUS), FRIGID DURORTHIDIC XERIC TORRIFLUVENTS
ALHARK	UT	FINE-LOAMY, MIXED, FRIGID XEROLLIC CALCIORTHIDS
ALICE	NE	COARSE-LOAMY, MIXED, MESIC ARIDIC HAPLUSTOLLS
ALICEL	OR	FINE-LOAMY, MIXED, MESIC PACHIC HAPLOXEROLLS
ALICIA	NM	FINE-SILTY, MIXED, MESIC USTOLLIC CAMBORTHIDS
ALIDA	IN	FINE-LOAMY, MIXED, MESIC AQUOLLIC HAPLUDALFS
ALIKCHI	OK	FINE-SILTY, SILICEOUS, THERMIC TYPIC GLOSSAQUALFS
ALINE	OK	SANDY, MIXED, THERMIC PSAMMENTIC PALEUSTALFS
ALKO	NV	LOAMY, MIXED, THERMIC, SHALLOW TYPIC DURORTHIDS
ALLAGASH	ME	COARSE-LOAMY OVER SANDY OR SANDY-SKELETAL, MIXED, FRIGID TYPIC HAPLORTHODS
ALLARD	NY	COARSE-SILTY OVER SANDY OR SANDY-SKELETAL, MIXED, MESIC TYPIC DYSTROCHREPTS
ALEGHENY	KY	FINE-LOAMY, MIXED, MESIC TYPIC HAPLUDULTS
ALLEMANDS	LA	CLAYEY, MONTMORILLONITIC, EUIC, THERMIC TERRIC MEDISAPRISTS
ALLEN	TN	FINE-LOAMY, SILICEOUS, THERMIC TYPIC PALEUDULTS
ALLENDALE	MI	SANDY OVER CLAYEY, MIXED, FRIGID AQUALFIC HAPLORTHODS
ALLENS PARK	CO	FINE-LOAMY, MIXED TYPIC EUTROBORALFS
ALLENTINE	MT	FINE, MONTMORILLONITIC, MESIC HAPLUSTOLLIC NATRARGIDS
ALLENWOOD	PA	FINE-LOAMY, MIXED, MESIC TYPIC HAPLUDULTS
ALLEY	NV	FINE-LOAMY, MIXED, MESIC DURIXEROLLIC HAPLARGIDS
ALLHANDS	ID	LOAMY-SKELETAL, MIXED, FRIGID, SHALLOW HAPLOXEROLLIC DURORTHIDS

TABLE 11/*Soil profile description from arroyo bank in alluvium derived from monzonite, classified Typic Torrifluvent; sandy, mixed, thermic (adapted from Gile and Grossman, 1979).*

Soil Classification: Typic Torrifluvent; sandy, mixed, thermic.

Soil: Vinton.

Soil Nos.: S59NMex-7-4.

Location: North bank of arroyo, SW 1/4 NW 1/4, Sec. 34, T21S, R3E, Dona Ana County, New Mexico.

Geomorphic Surface: Organ.

Landform: Ridge sloping 3 percent to the west.   Elevation: 4,735 feet.

Parent Material: Organ alluvium derived from monzonite. Sediments from 0 to 58 cm are Organ; from 58 to 135 cm probably early Organ; from 135 to 213 cm, Jornada II.

Vegetation: Dominantly snakeweed, few clumps black grama, fluffgrass, few Mormon tea, *Yucca elata*, very few creosotebush.

Collected by: J. S. Allen, L. H. Gile, R. B. Grossman, and R. V. Ruhe, May 12, 1959.

Described by: L. H. Gile and R. B. Grossman.

Soil Surface: Discontinuous layer of fine monozonite gravel, mainly less than 1/4 inch in diameter.

A2   11258 0 to 5 cm. Brown (10YR 5/3 dry) or dark brown (10YR 3/3 moist) coarse loamy sand; weak thin and medium platy; soft; few roots; noncalcareous; abrupt smooth boundary.

B2t   5 to 23 cm. Brown (7.5YR 5/4 dry) or dark brown (7.5YR 3.5/4 moist) coarse sandy loam; massive to single grain; soft; roots few to common; noncalcareous; abrupt wavy boundary.

C1ca   11260 23 to 38 cm. Brown (10YR 5/3 dry) or dark yellowish-brown (10YR 3/4 moist) coarse sandy loam; massive; soft to slightly hard; roots few to common; effervesces weakly and strongly; abrupt wavy boundary.

IIC2ca   11261 38 to 58 cm. Brown (10YR to 7.5YR 5/3 dry) or dark brown (10YR to 7.5YR 3.5/3 moist) fine gravelly sand; single grain to massive; loose to soft; few roots; thin discontinuous carbonate coatings on pebbles; effervesces strongly; abrupt wavy boundary.

IIIB1b   11262 58 to 66 cm. Brown (10YR 5/3 dry) or dark yellowish-brown (10YR 3.5/4 moist) coarse sandy loam; massive; slightly hard; few roots; effervesces weakly; clear wavy boundary.

IVB2cab   11263 66 to 74 cm. Brown (7.5YR 5/4 dry) or dark brown (7.5YR 3.5/4 moist) fine gravelly coarse sandy loam; massive; slightly hard; few roots; thin discontinuous carbonate coatings on some pebbles; generally effervesces weakly and strongly with a few parts noncalcareous; clear wavy boundary.

IVC1cab   11264 74 to 97 cm. Brown (7.5YR 5/4 dry) or dark brown (7.5YR 3.5/4 moist) fine very gravelly loamy coarse sand; 80 percent single grain, 20 percent massive; loose and soft; few roots; undersides of larger pebbles coated with carbonate; effervesces weakly and strongly; clear wavy boundary.

VC2cab   11265 97 to 135 cm. Brown (7.5YR 5/4 dry) or dark brown (7.5YR 3.5/4 moist) coarse sandy loam; massive; friable; very few roots; few carbonate filaments; thin, discontinuous carbonate coatings on pebbles and on walls of larger voids; few weak dull films on larger void surfaces; insect holes up to 1/2 inch in diameter are common and have thin coatings of fine material, the same color as rest of horizon; effervesces weakly and strongly; clear wavy boundary.

VIB1cab2   11266 135 to 160 cm. Brown (7.5YR to 5YR 5/4 dry) or dark brown (7.5YR 3.5/3 moist) coarse sandy loam; massive; friable; very few roots; reflective surfaces on sand grains common; few carbonate filaments; effervesces weakly; clear to gradual boundary.

VIB21cab2   11267 160 to 198 cm. Reddish-brown (5YR 5/4 dry) or dark reddish-brown (5YR 3.5/4 moist) sandy clay loam; weak medium subangular blocky; firm; few roots; few reflective surfaces on weak ped faces and on walls of noncapillary size voids; few carbonate filaments; effervesces weakly and strongly; clear smooth boundary.

VIB22cab2   11268 198 to 213 cm. Fifty percent reddish-brown (5YR 5/4 dry, 4/4 moist) and 50 percent pinkish-white (5YR 8/2 dry) or light reddish-brown (5YR 6/3 moist), the latter relatively rich in carbonate, sandy clay loam; massive; firm; very few roots; carbonate occurs fairly continuously throughout; effervesces strongly.

Remarks: The VC2cab horizon (97 to 135 cm.) may be a buried B, related to the IIB2cab horizon of S59NMex-7-3.

Mineralogy (Methods 7A2, 7A3): The Beltsville laboratory determined the clay mineralogy of the C1ca horizon (23-38 cm). A poorly organized montmorillonite-vermiculite intergrade is abundant; a small amount of mica and 15 percent kaolinite are present.

TABLE 12/*Soil Taxonomy characterization data for the Vinton soil in Dona Ana County, New Mexico (adapted from Gile and Grossman, 1979).*

SOIL CLASSIFICATION: Typic Torrifluvent; sandy, mixed, thermic

U. S. DEPARTMENT OF AGRICULTURE
SOIL CONSERVATION SERVICE

SOIL __Vinton__   SOIL Nos. __S59NMex-7-4__   LOCATION __Dona Ana County, New Mexico__

SOIL SURVEY LABORATORY __Lincoln, Nebraska__   LAB. Nos. __11258-11268, 72L575__

GENERAL METHODS: 1A1, 1B1a, 2A1, 2B, a/

Depth (cm.)	Horizon	Total Sand (2–0.05)	Total Silt (0.05–0.002)	Total Clay (< 0.002)	Very coarse (2–1)	Coarse (1–0.5)	Medium (0.5–0.25)	Fine (0.25–0.1)	Very fine (0.1–0.05)	Silt 0.05–0.02	Int. III (0.02–0.002)	Int. II (0.2–0.02)	(2–0.1)	Coarse fragments Vol. 250-2 %<250	Wt. 76-2 %<76	Wt. 19-2 %<19
								Carbonate Removed								
0-5	A2	78.0	16.3	5.7	15.2	15.8	8.4	19.1	19.5	12.3	4.0	43.0	58.5	5	10	10
5-23	B2t	76.4	15.6	8.0	19.1	16.5	7.9	15.7	17.2	11.9	3.7	38.1	59.2	10	16	16
23-38	C1ca	78.9	14.2	6.9	22.2	17.5	7.8	14.6	16.8	10.6	3.6	35.8	62.1	10	15	15
38-58	IIC2ca	87.4	8.0	4.6	28.8	21.9	11.0	15.8	9.9	5.8	2.2	23.5	77.5	15	24	24
58-66	IIIB1b	75.1	16.6	8.3	15.5	14.7	7.8	16.7	20.4	12.7	3.9	42.8	54.7	10	20	20
66-74	IVB2cab	77.9	12.9	9.2	24.8	16.3	7.6	14.7	14.5	9.3	3.6	32.0	63.4	10	22	22
74-97	IVC1cab	79.9	13.3	6.8	25.6	19.2	8.0	13.6	13.5	9.8	3.5	30.5	66.4	15	28	28
97-135	VC2cab	71.6	18.5	9.9	15.4	13.0	7.7	16.8	18.7	13.8	4.7	41.9	52.9	10	22	22
135-160	VIB1cab2	74.2	14.4	11.4	16.6	15.0	8.0	16.9	17.7	10.1	4.3	37.4	56.5	10	15	15
160-198	VIB2ab2	60.6	20.4	19.0	14.7	11.0	6.2	13.2	15.5	14.0	6.4	37.3	45.1	15	23	23
198-213	VIB22cab2	69.4	15.8	14.8	19.1	14.4	7.3	14.9	13.7	10.5	5.3	32.7	55.7	10	22	22

Depth (cm.)	6A1a Organic carbon b/, g/ Pct.	6B1a Nitrogen b/ Pct.	C/N	6C1a Ext. Iron as Fe Pct.	Carbonate as CaCO₃ <2mm. d/ Pct.	Carbonate as CaCO₃ <2mm. e/ Pct.	Bulk density f/ g/cc	Bulk density g/cc	CEC NH₄OAc me./100g.	Water content Pct.	Water content Pct.	Water content Pct.	Composition Whole Material NONCARBONATE >2mm. Pct.	Sand Pct.	Silt Pct.	Clay Pct.	Carbonate as CaCO₃ Pct.
0-5	0.19	0.026	7	1.1	0.3		1.4		6.0				10	71	14	5	tr
5-23	0.22	0.022	10	0.9	-		1.4		6.5				16	64	13	7	-
23-38	0.25	0.029	9	0.9	2	4.4	1.4		6.7				15	65	12	6	2
38-58	0.18	0.015		1.0	1	5.2	1.4		4.8				24	65	6	4	1
58-66	0.19	0.018		0.9	1	1.5	1.4		7.1				20	60	13	6	1
66-74	0.19	0.020		0.8	1	3.0	1.4		7.2				22	60	10	7	1
74-97	0.13	0.016		0.8	3	4.4	1.4		6.4				27	57	9	5	2
97-135	0.14	0.015		0.9	1	1.3	1.4		8.6				22	54	15	8	1
135-160	0.13	0.015		1.0	1	2.8	1.4		8.5				15	63	12	9	1
160-198	0.15	0.019		1.1	1		1.4		12.2				23	47	15	14	1
198-213	0.11	0.013		0.8	13		1.4		11.0				20	48	11	10	11

a/ The < 2 mm. was removed from the > 2 mm. by gentle agitation in water. This < 2 mm. was added to the < 2 mm. obtained by dry sieving. Carbonate was removed from the composited sample by Method 1B3. Determinations were made on and reported for the composite sample so treated unless otherwise indicated.

b/ Determination on sample treated by Method 1B3 to remove carbonate; values expressed on a carbonate-containing basis.

c/ Inclusive of coarse fragments, carbonate, and gypsum.

d/ Determination on whole material by Methods 6E1a and 6E1c and calculated as a percentage of the < 2 mm.

e/ < 2 mm. obtained by washing the > 2 mm. gently in water; Method 6E1c.

f/ Bulk density assumed for moist fine-earth fabric for calculations.

g/ 2.3 kg/m² to 97 cm (Method 6A).

h/ Volume on carbonate-containing basis; weight on carbonate-free basis.

TABLE 13/*Soil profile description of Arenic Fragiudult, clayey, kaolinitic, isohyperthermic soil at Campo Alegre in Brazil (adapted from Camargo and Beinroth, 1978).*

Profile: ISCW-BR 20.   Described and Sampled Feb. 17, 1977.

Classification: Podzólico vermelho-amarelo distrófico, argila de atividade baixa frágico A proeminente textura arenosa/arilosa fase floresta tropical subperenifólia relevo plano (red-yellow podzolic dystrophic, low clay activity, fragic, prominent A horizon, sandy/clayey, semievergreen tropical forest level phase).
Arenic Fragiudult; clayey, kaolinitic, isohyperthermic.
Dystric Planosol.
Sol ferrallitique; fortement désaturé, lessivé, podzolisé, dérivé de formation Barreiras.

Location: Campo Alegre, Al. km 163 of the BR-101 highway, São Sebastião farm; 9°50'00" S, 36°12'00" W.

Topographic Position: Trench on level top of low plateau (tableland), under sugarcane field.

Primary Vegetation: Semievergreen tropical forest.

Geology and Parent Material: Sandy and sandy clay sediments, Barreiras Group, Tertiary; weathered sediments.

Drainage: Moderately well drained.

Present Land Use: Sugarcane cultivated since 1967.

	Jan	Feb	Mar	Apr	May	Jun	Jul	Aug
T (°C)	26.0	26.0	26.0	25.5	25.0	24.0	23.5	23.0
P (mm)	50	60	110	200	230	200	200	135

	Sept	Oct	Nov	Dec		
T (°C)	23.5	24.0	25.0	25.5	Mean	24.8
P (mm)	70	50	45	35	Total	1385

Isohyperthermic   Ustic/udic

Ap   0–45 cm, very dark grayish brown (10YR 3/2, moist), dark grayish brown (10YR 4/2, dry); loamy sand; weak fine to medium granular and subangular blocky; many very fine and fine, and few medium and coarse pores; soft, very friable, nonplastic and nonsticky; gradual and smooth boundary.

A2   45–70 cm, brown (10YR 4/3, moist); pale brown (10YR 6/3, dry); loamy sand; weak fine subangular blocky and fine medium granular; many very fine and fine, and few medium and coarse pores; slightly hard, very friable, nonplastic and nonsticky; gradual and smooth boundary.

A3   70–95 cm, brown (10YR 5/3, moist), light brownish gray (10YR 6/2, dry); sandy loam; weak fine angular and subangular blocky; common very fine and fine, and few coarse pores; hard, friable, nonplastic and nonsticky; clear and wavy boundary.

B1t   95–112 cm, brown (10YR 5/3, moist); sandy clay; weak fine angular and subangular blocky; common very fine and fine, and few medium pores; very hard, firm and friable, plastic and sticky.

B21tx   112–133 cm, mixture of light yellowish brown (10YR 6/4, moist) and light brownish gray (10YR 6/2, moist) colors; clay; weak fine platy and weak fine to medium angular blocky; few very fine and fine, and few medium pores; extremely hard, firm, slightly plastic and slightly sticky; abrupt and wavy boundary (15–25 cm).

B22tx   133–145 cm, mixture of brownish yellow (10YR 6/6, moist) and brown (10YR 5/3, moist) colors; clay; weak fine platy and weak fine angular blocky; few very fine and fine; extremely hard, very firm, slightly plastic and slightly sticky; gradual and smooth boundary.

B23t   145–200 cm⁺, light yellowish brown (10YR 6/4, moist), few fine and medium distinct mottles of yellowish brown (10YR 5/8, moist); clay; weak fine subangular blocky; few very fine and fine pores; extremely hard, firm, slightly plastic and nonsticky.

Remarks: Abundant roots in Ap, common in A2 and A3 and upper part of B1t, few in lower part of B1t and very few downward.
Plaquic? layer of Fe_2O_3 with average thickness of 1 cm, predominantly with red color (2.5YR 4/8, moist) in B22tx.
Portions of darker material intermingled in B21tx, B22tx, and B23t, being mostly horizontal in B21tx, seemingly after the direction of roots or other biological activity.
Reticular pattern of colors in B21tx being less evident in B22tx.
Ant and termite activity from the surface down to B21tx.
Profile dry.

TABLE 14/*Soil classification laboratory data for Arenic Fragiudult, clayey, koalinitic, isohyperthermic soil at Campo Alegre in Brazil (adapted from Camargo and Beinroth, 1978).*

PROFILE Nº ISCW-BR 20
SAMPLE Nº 77.0780/86   SNLCS

HORIZON	DEPTH cm	GRAVEL >20 mm %	GRAVEL 20-2mm %	FINE EARTH <2mm %	CORS 2-.20 mm	FNES .20-.05 mm	SILT .05-.002 mm	CLAY <.002 mm	WATER DISP CLAY %	FLOC DEGREE %	SILT/CLAY
Ap	0-45	-	tr	100	66	22	2	10	5	50	0.20
A2	-70	-	tr	100	64	23	3	10	7	30	0.30
A3	-95	-	1	99	51	24	5	20	13	35	0.25
B1t	-112	-	1	99	36	22	4	38	31	18	0.11
B21tx	-133	-	tr	100	28	14	6	52	30	42	0.12
B22tx	-145	-	1	99	27	14	7	52	35	33	0.13
B23t	-200+	-	tr	100	26	14	7	53	5	91	0.13

pH H_2O	KCL N	Ca++	Mg++	K+	Na+	SUM EXTR	Al+++	H+	CAT EXCH mE/100g	BASE SAT %	$\frac{100 \cdot Al^{+++}}{Al^{+++}+S}$
5.2	4.2	1.0	0.2	0.07	0.05	1.3	0.2	2.8	4.3	30	13
5.2	4.2	0.8		0.04	0.03	0.9	0.2	2.6	3.7	24	18
4.6	4.0	0.6		0.03	0.03	0.7	0.6	2.4	3.7	19	46
4.8	4.1	0.6		0.03	0.04	0.7	0.6	2.2	3.5	20	46
5.1	4.5	0.9		0.03	0.03	1.0	0.5	2.3	3.8	26	33
5.1	4.5	1.0		0.03	0.04	1.1	0.4	2.1	3.6	31	27
5.2	4.5	1.0		0.03	0.03	1.1	0.4	2.0	3.5	31	27

ORG C %	N %	C/N	SiO_2	Al_2O_3	Fe_2O_3	TiO_2	SiO_2/Al_2O_3	SiO_2/R_2O_3	Al_2O_3/Fe_2O_3	AVLB PHOS ppm
0.63	0.06	11	4.1	3.3	0.5	0.41	2.11	1.92	10.45	1
0.38	0.04	10	4.5	3.7	0.5	0.45	2.07	1.90	11.71	1
0.39	0.04	10	8.8	7.6	0.7	0.69	1.97	1.86	16.93	1
0.37	0.04	9	16.1	14.4	1.2	1.01	1.90	1.80	18.83	1
0.38	0.04	10	22.0	20.2	1.9	1.33	1.85	1.75	16.64	1
0.28	0.04	7	22.6	21.1	2.0	1.28	1.82	1.72	16.55	1
0.23	0.03	7	23.4	21.5	2.2	1.29	1.85	1.74	15.28	1

Clay B/A - 3.8   Weighted - 3.8

Computerized groupings of soils

PURPOSE This exercise illustrates techniques by which soils are grouped for practical purposes, on the basis of the characteristics of the soils. Previous exercises have been involved with the language (terms) of the soil survey, and this exercise and those following deal with the applications of the information. Computers are particularly well adapted to the use of soil information, so this exercise will emphasize the current state of computer technology. Of course, computers are only a tool to facilitate efficiency in soil survey interpretations; groupings of soils can be accomplished manually as well as by computer manipulations. Pages 62–90 of the textbook should be studied in preparation for this exercise.

GATHER INFORMATION Gathering information is the first step in making groupings of soils for practical purposes. Accessory information (from agronomy, conservation, economics, engineering, forestry, planning, etc.) is fully as important at this stage as soil profile descriptions, soil maps, and soils data. Coordinations and correlations of interdisciplinary data mark the accomplishments of soil survey interpretations.

Assemble all the soils information you have accumulated from your study area, including soil profile descriptions, soil map, laboratory data, and Soil Taxonomy classifications. List all the soil properties that you think are important for crop production (see page 67 of the textbook). Then tabulate the pH values and other lab data that you obtained for your soils. Combine the chemical and physical data to make recommendations about uses of the soils in your study area. If the soil is poorly drained, specify that drainage is needed for improved crop growth. If the pH is low, you can recommend liming to raise the pH and the cation exchange to plant roots. Contour cultivation or minimum tillage practices can be specified for sloping soils to reduce erosion.

Table 15 provides a computer illustration of input information that accompanies a soil sample (taken from the topsoil) for fertility analyses in the laboratory. Soil survey information includes soil name and map symbol, depth, drainage, texture, and topography. Information about past cropping and future yield goals and expectations are included. Soil fertility analyses from the laboratory are combined with the information in Table 15 to provide the data base for making recommendations to the farmer and grower (Hahn et al., 1978; Olson et al., 1982). All this information is considered by Cooperative Extension Agents, Soil Conservationists, and others when making recommendations to farmers—and by the farmers themselves when making their management decisions. Programs are written for the computer that help to make recommendations more systematic and standardized. Computers also provide a "storage bank" for information (see Table 8), so that data about fields and sites can be accumulated over long periods of time for historical perspectives of soil fertility changes and land-use shifts.

SOIL SURVEY INTERPRETATIONS Table 16 is the computer input form (SCS Form 5) for soil survey interpretations in the United States; other countries have similar input forms for making ratings of soils for various uses. Table 17 provides brief instructions for filling out Form 5. Table 18 provides an example of the computer output form from the inputs of Tables 16 and 17. Table 19 lists some of the standardized terms and phrases used in making soil survey interpretations. Tables 20 to 27 provide some examples of criteria by which soil properties are evaluated and soils are rated for the various uses. The rating tables are taken from Part II, Section 400, on "Application of Soil Survey Information" from the National Soils Handbook (Soil Survey Staff, 1978).

Each teacher and class should procure a copy of the part of the National Soils Handbook relevant to application of soil survey information from the offices of the Soil Conservation Service or the Agricultural Experiment Station soil survey representatives of the Cooperative Soil Survey. Part II, Section 400 (Soil Survey Staff, 1978), has about 100 pages of detailed procedures and criteria for using soil characteristics in application of soil survey information, from which some examples are illustrated in Tables 16 to 27.

Standardized soil ratings into three categories essentially group the "best" soils for a specific use into categories of "slight" limitations or a "good" performance category. The "worst" soils for a certain use are put into a "severe" limitation group or a "poor" category. All other soils are rated "moderate" or "fair" for that use. Decisions for rating are based on criteria such as those illustrated in Tables 20 to 27, and one severe rating of only one soil characteristic will put that soil in the most limiting category, regardless of the ratings of the other characteristics. Of course, this rating process is to provide general guidelines only; rating criteria can be revised and modified to fit local conditions. "Severe" or "poor" ratings are of limited utility to the users of soil information. What people need most are statements about problems to be overcome in each soil, and specifications about economic inputs that need to be made to overcome the limitations and achieve the fullest potential of each soil area.

According to the National Soils Handbook (Soil Survey Staff, 1978):

Slight is the rating given soils that have properties favorable for the use. The degree of limitation is minor and can be overcome easily. Good performance and low maintenance can be expected.

Moderate is the rating given soils that have properties moderately favorable for the use. This degree of limitation can be overcome or modified by special planning, design, or maintenance. During some part of the year, the expected performance of the structure or other planned use is somewhat less desirable than for soils rated *slight*. Some soils rated *moderate* require such treatment as artificial drainage, control of runoff to reduce erosion, extended septic tank absorption fields, extra excavation, or some modification of certain features through manipulation of the soil. For these soils, modification is needed for those construction plans generally used for soils of *slight* limitations. Modification may include specially designed foundations, extra reinforcement of structures, sump pumps, and the like.

Severe is the rating given soils that have one or more properties unfavorable for the rated use, such as steep slopes, bedrock near the surface, flooding, high shrink–swell potential, a seasonal high water table, or low strength. This degree of limitation generally requires major soil reclamation, special design, or intensive maintenance. Some of these soils, however, can be improved by reducing or removing the soil feature that limits use, but, in most situations, it is difficult and costly to alter

the soil or to design a structure so as to compensate for a severe degree of limitation.

Students should fill out the blanks in the SCS Form 5 (Table 16) for the soils in their study area that they described, mapped, and analyzed in the laboratory. For the Form 5 computer entry, ratings must be according to the strict standards outlined in the National Soils Handbook (Soil Survey Staff, 1978). These ratings in the computer are used for rating each soil map unit in the tables produced for published soil survey reports. Other Form 5 ratings should be checked with ratings for similar soils to be sure that the student ratings are consistent and appropriate. However, students should be encouraged to use initiative and imagination in making application of the ratings. If most areas in a county are rated "severe" for septic tank absorption fields, for example, the users of soils information must be informed more fully about the variable soil properties, and what must be done to overcome the limitations.

One example of elaboration of the three-class rating system is illustrated on pages 66–67 of the textbook. Page 66 has a simple three-class rating system that is useful for a first approximation, and page 67 has a five-class system that is more complex but more useful in many situations. Similarly, all the guidelines in the National Soils Handbook can be modified and refined to fit better into local situations and environments. Table 28 is an illustration of modification of the table on page 67 of the textbook (for general farming in New York State) to fit soils used for rain-fed and irrigated paddy rice in Southeast Asia. Obviously, many soil characteristics have opposite effects for corn versus rice, but the same soil properties are described and mapped in New York and Thailand and can be rated for any intended use with any assumed criteria. As more data become available, the ratings can become more refined, precise, and elaborate.

The final step in making soil ratings is to transfer the information to maps for practical uses. Figures 15 to 21 illustrate some of the results for the area surrounding Lagol about 50 miles northwest of Los Angeles (see pages 71–77 of the textbook). Figure 15 is the detailed soil map of the area, and Figures 17 to 21 are interpretive maps made to show hydrologic soil groups, soil erosion potential, shrink–swell potential, suitability for farming, and soil limitations for avocado root rot. Figure 16 is an overrun soil map without the air photo background, which is especially valuable for coloring the various patterns of the different soils. Green is often used for the best areas, yellow for moderate, and red (danger) for areas with severe limitations. Hand coloring of these maps, of course, is very laborious and time consuming. Most individual student projects should concentrate on a relatively limited area, so that excessive amounts of time are not consumed by the hand coloring.

Computers, of course, provide a much more efficient method of producing colored and digitized interpretive soils maps. Criteria such as those contained in Tables 20 to 27 can be entered into the computer (together with the soil map) for rating each soil in a survey area (Soil Survey Staff, 1982), and then the computer can quickly produce the maps required. Figures 22 to 29 illustrate interpretive soils maps made by computer for Benton County, Arkansas (Soil Survey Staff, 1982). These maps show depth to bedrock; suitability for agriculture, orchards and vineyards, truck farming, and pasture; limitations for light industry, sanitary landfills (trench type), and limitations for parks and playgrounds. The grid cell size in these maps is about 10 acres, but cells of any size can be used. Cell size generally should be smaller than the minimum-size delineations on the soil map, and cells of 1 hectare or 1 acre will show most of the detail of most soil maps made at a scale of 4 inches to 1 mile. Some computers will reproduce soil maps by a line-segment system that approximates the true soil boundaries of the original soil map.

As part of this exercise, students should computerize soils information and reproduce interpretive soils maps by computer. Many agencies and offices are already using various computer techniques to accomplish these objectives, and the class teachings and student projects should be coordinated to local programs under way. As McRae and Burnham (1981) had stated: "The future of land evaluation (soil survey interpretations) lies in storing and processing data with computers."

REFERENCES

Edwards, R. D., D. F. Rabey, and R. W. Kover. 1970. Soil survey of the Ventura Area, California. University of California Agriculture Experiment Station and Soil Conservation Service, U.S. Dept. of Agriculture, U.S. Government Printing Office, Washington, DC. 151 pages and 50 soil map sheets.

Hahn, R. R. et al. 1978. Cornell field crops handbook. New York State College of Agriculture and Life Sciences, Cornell University, Ithaca, NY. 160 pages.

McRae, S. G. and C. P. Burnham. 1981. Land evaluation. Clarendon Press, Oxford. 239 pages.

Olson, G. W. (Editor) et al. 1982. 1982 Cornell recommends for field crops. New York State College of Agriculture and Life Sciences, Cornell University, Ithaca, NY. 56 pages.

Soil Survey Staff. 1978. National soils handbook. Part II, Section 400, Application of soil survey information, Procedure guide. Soil Conservation Service, U.S. Dept. of Agriculture, Washington, DC. Reproduced as Cornell Agronomy Mimeo 82-15, Cornell University, Ithaca, NY. About 100 pages.

Soil Survey Staff. 1982. Computer generated resource data and colored map display. Soil Conservation Service, U.S. Dept. of Agriculture, Fort Worth, TX. Reproduced as Cornell Agronomy Mimeo 82-14, Cornell University, Ithaca, NY. 27 pages.

FIGURE 15/*Soil Map Sheet 27 of area surrounding Lagol about 50 miles northwest of Los Angeles, California. Scale of original map 1:24,000 (adapted from Edwards et al., 1970).*

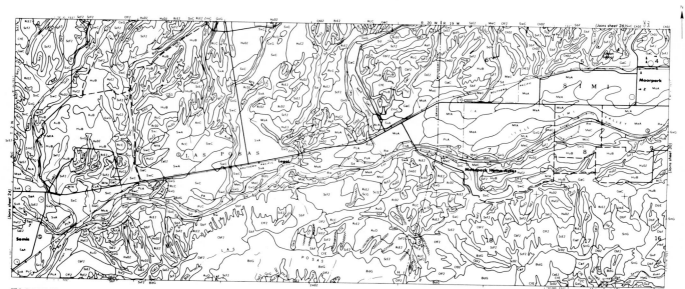

FIGURE 16/*Overrun soil map (without air photo background) made from Soil Map Sheet 27 of area surrounding Lagol, California. Scale of original map 1:24,000 (adapted from Edwards et al., 1970).*

FIGURE 17/*Hydrologic soil groups in areas of Soil Map Sheet 27 of area surrounding Lagol, California. Scale of original map 1:24,000 (adapted from Edwards et al., 1970).*

FIGURE 18/*Soil erosion potential in areas of Soil Map Sheet 27 of area surrounding Lagol, California. Scale of original map 1:24,000 (adapted from Edwards et al., 1970).*

FIGURE 19/*Shrink–swell potential of the soils surrounding Lagol, California. Scale of original map 1:24,000 (adapted from Edwards et al., 1970).*

FIGURE 20/*Soil suitability for farming in the area surrounding Lagol, California. Scale of original map 1:24,000 (adapted from Edwards et al., 1970).*

FIGURE 21/*Soil limitations for avocado root rot in the area surrounding Lagol, California. Scale of original map 1:24,000 (adapted from Edwards et al., 1970).*

40

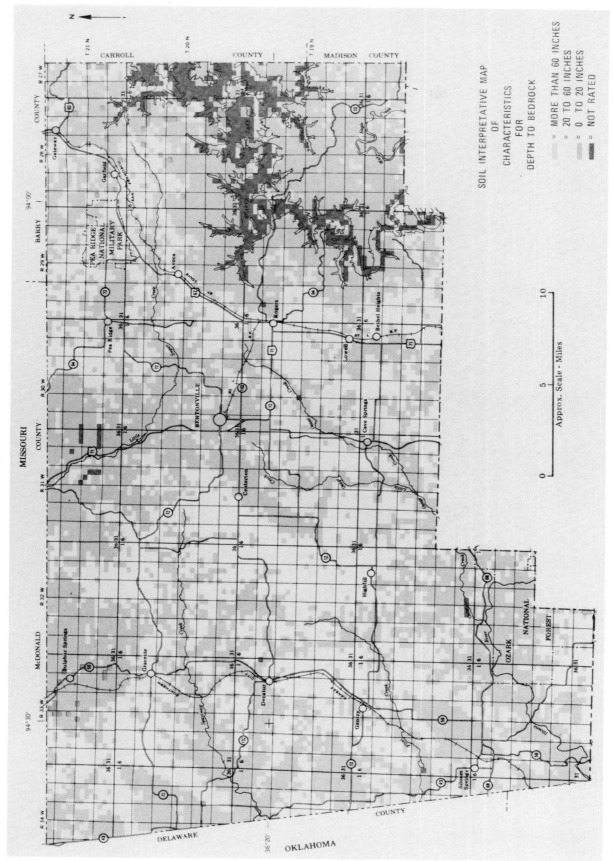

FIGURE 22/Computer-generated soil interpretive map of characteristics for depth to bedrock in Benton County, Arkansas (adapted from Soil Survey Staff, 1982).

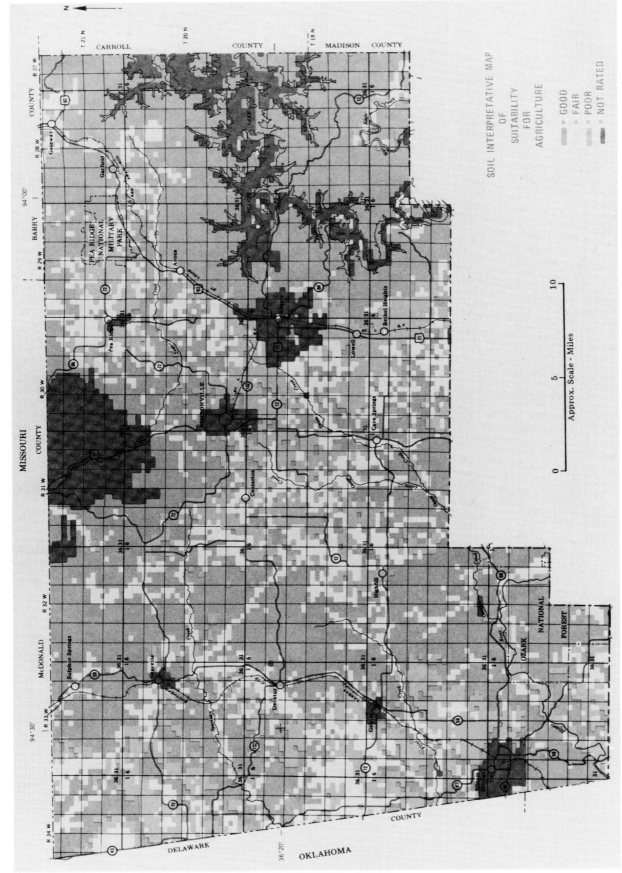

FIGURE 23 /Computer-generated soil interpretive map of suitability for agriculture in Benton County, Arkansas (adapted from Soil Survey Staff, 1982).

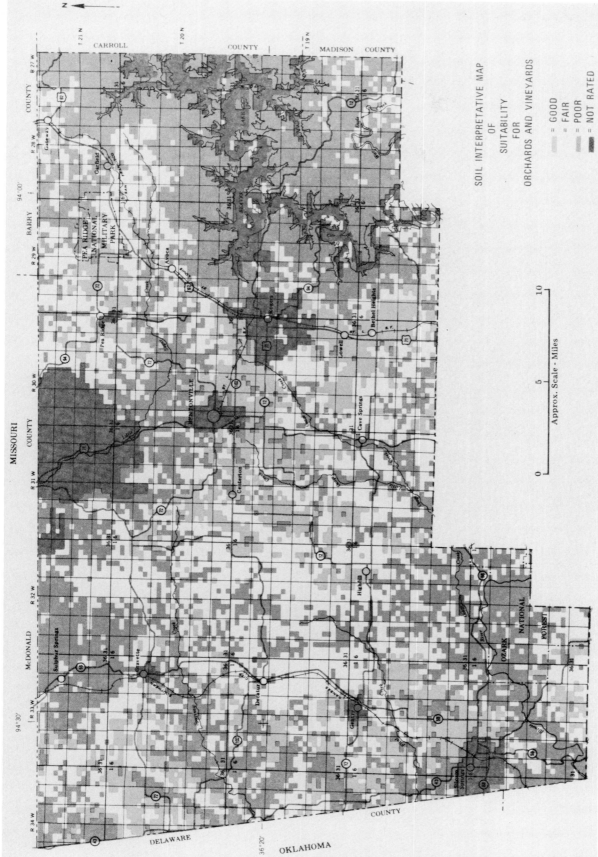

FIGURE 24/*Computer-generated soil interpretive map of suitability for orchards and vineyards in Benton County, Arkansas (adpated from Soil Survey Staff, 1982).*

43

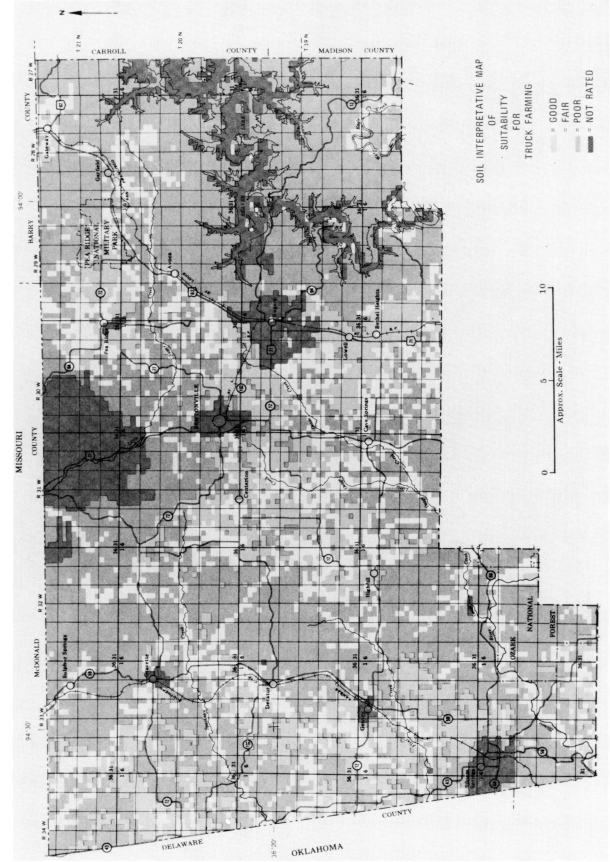

FIGURE 25/*Computer-generated soil interpretive map of suitability for truck farming in Benton County, Arkansas (adapted from Soil Survey Staff, 1982).*

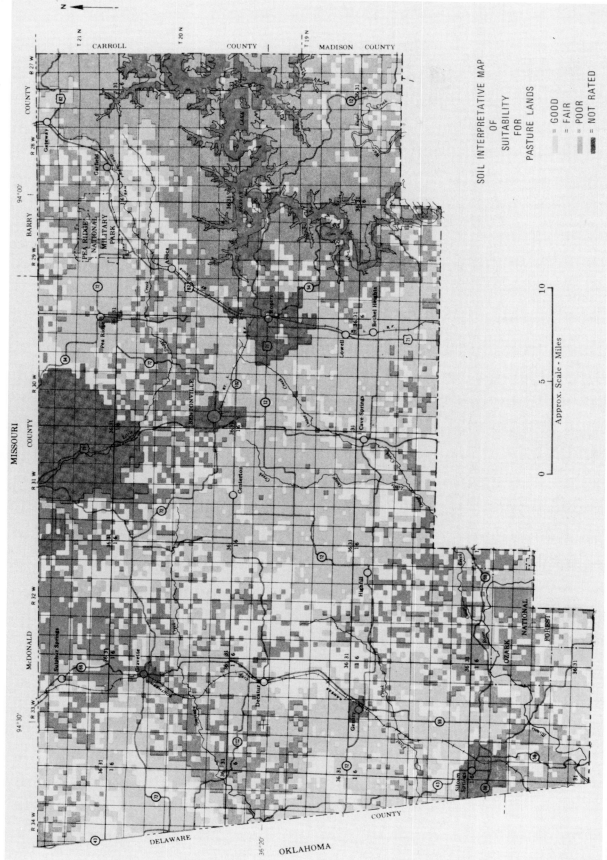

FIGURE 26/*Computer-generated soil interpretive map of suitability for pasturelands in Benton County, Arkansas (adapted from Soil Survey Staff, 1982).*

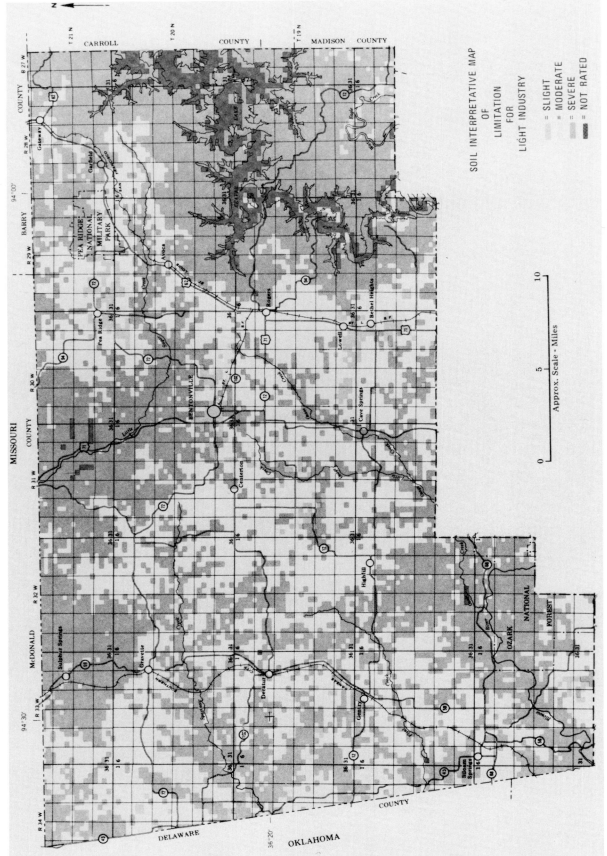

FIGURE 27/*Computer-generated soil interpretive map of limitation for light industry in Benton County, Arkansas (adapted from Soil Survey Staff, 1982).*

SOIL INTERPRETATIVE MAP
OF
LIMITATION
FOR
LIGHT INDUSTRY

= SLIGHT
= MODERATE
= SEVERE
= NOT RATED

Approx. Scale - Miles

0   5   10

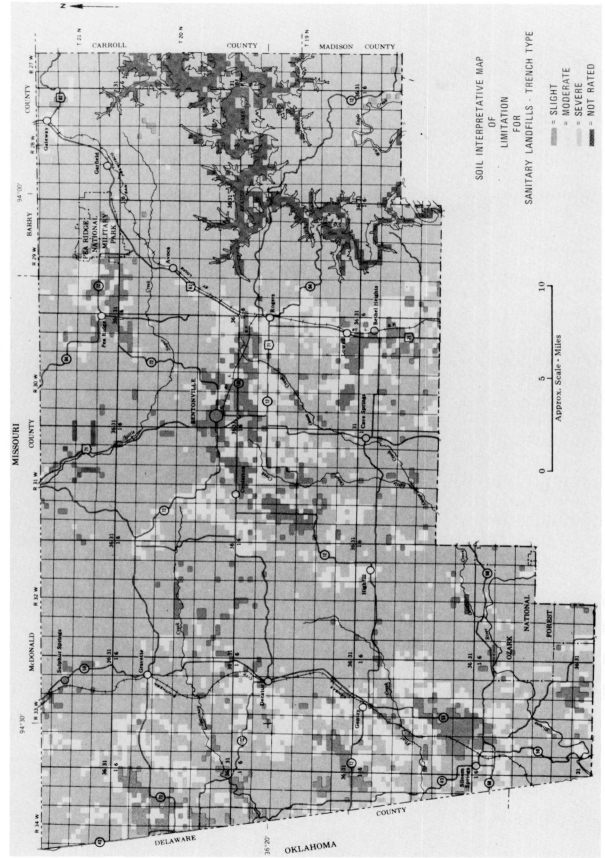

FIGURE 28/*Computer-generated soil interpretive map of limitation for sanitary landfills (trench type) in Benton County, Arkansas (adapted from Soil Survey Staff, 1982).*

47

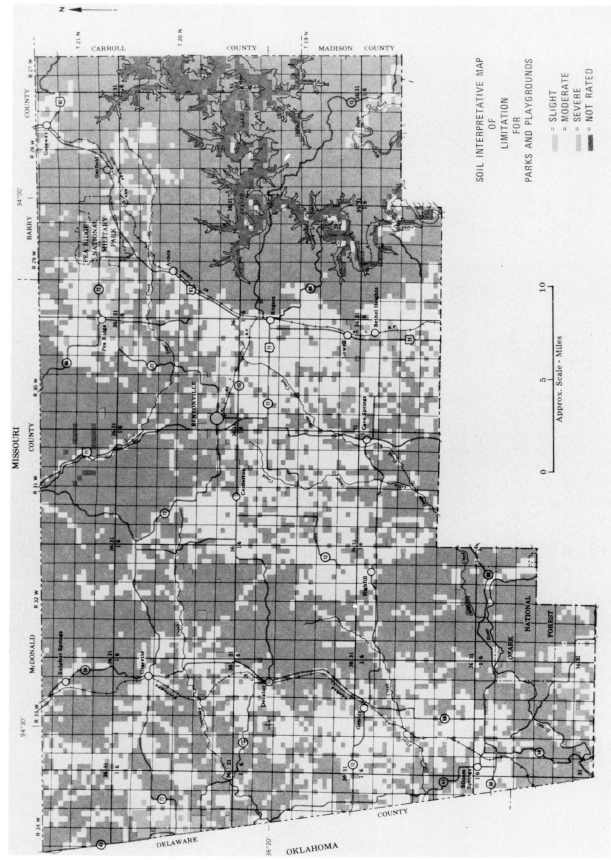

FIGURE 29 / *Computer-generated soil interpretive map of limitations for parks and playgrounds in Benton County, Arkansas (adapted from Soil Survey Staff, 1982).*

TABLE 15/Computer input form used in New York State to provide information to accompany sample sent to the laboratory for soil fertility testing.

SOIL TESTING SERVICE · DEPARTMENT OF AGRONOMY
COLLEGE OF AGRICULTURE AND LIFE SCIENCES · CORNELL UNIVERSITY
BRADFIELD HALL, ITHACA, NEW YORK 14853

OFFICE USE

INFORMATION SHEET FOR **FIELD AND VEGETABLE CROPS** USE
OTHER INFORMATION SHEETS FOR FRUITS, GARDENS, ORNAMENTALS AND TURF

SEND THIS COPY TO LABORATORY WITH SAMPLE·YOUR REPORT CANNOT BE COMPLETED WITHOUT IT

1. NAMES AND ADDRESSES

GROWER'S NAME

STREET ADDRESS

CITY | STATE | ZIP

TELEPHONE NO

COMMERCIAL REPRESENTATIVE NAME

COMMERCIAL FIRM NAME

STREET ADDRESS

CITY | STATE | ZIP

LOCATION OF FIELD

COUNTY

TOWNSHIP

County to receive results if different from above

2. SAMPLE AND FIELD IDENTIFICATION

SAMPLE BAG NO.
Copy 5 digit number from envelope on sample bag

Field Name or Number (up to 15 digits)

IS FIELD ARTIFICIALLY DRAINED? (X) ☐ YES ☐ NO

Acres (to the nearest acre)

MO-DAY-YR
Date Sampled

☐ Yes ☐ No
ASCS Copy? (X)

3. SOILS IDENTIFICATION & INFORMATION

SOIL NAME

Obtain the soil name occupying the largest area of the field from a soil survey or conservation plan

Soil name, e.g. Mardin silt loam

MAP UNIT SYMBOL
e.g. MaB

ORGANIC SOILS ONLY
Depth of muck or peat soils (X)

☐ Deep (>24 in)

☐ Shallow (<24 in)

☐ Shallow mixed with marl

☐ Shallow mixed with mineral soil

PLOW DEPTH (X)

☐ No till

☐ 1 - 7 in

☐ 7 - 9 in

☐ 9 + in

COMPLETE THIS SOIL ASSOCIATION SECTION IF SOIL NAME AND MAP UNIT SYMBOL ARE NOT GIVEN

SOIL ASSOCIATION SYMBOL

Obtain soil association symbol from the county soil association map (Obtain map from Cooperative Extension)

_____ symbol (i.e., E4-2)

SOIL DRAINAGE (X)

☐ Well

☐ Moderately well

☐ Somewhat poorly

☐ Poorly

TEXTURE (X)

☐ Sandy

☐ Loamy

☐ Silty

☐ Clayey

TOPOGRAPHY (X)

☐ Hilltop

☐ Hillside

☐ Valley floor

☐ Central Plains

4. CROPS GROWN

Place crop codes from the back of second sheet in blocks below

PAST CROPS

☐☐☐ LAST YEAR'S CROP

☐☐☐ 2 YEARS AGO'S CROP

☐☐☐ 3 YEARS AGO'S CROP

CROPS TO BE GROWN

☐☐☐ NEXT YEAR'S CROP   YEAR 19 ___

☐☐☐ 2ND YEAR'S CROP   19 ___

☐☐☐ 3RD YEAR'S CROP   19 ___

OTHER_____

LEGUME CONTENT

ESTIMATE PERCENT LEGUME in existing hay or pasture; or if plowed, in the last sod.

☐ None (100% grass)

☐ 1-25% legume

☐ 25-50% legume

☐ 50-100% legume

PLOWED SOD

Give Year Sod was Plowed (e.g., 1975)

19_____

5. COVER CROP

If cover crops are grown- indicate when planting (X)

☐ Before next year's crop

☐ Before 2nd year's crop

☐ Before 3rd year's crop

6. MANURE USE

RATE (Tons/Acre)

_____ Last year's crop

_____ This year's crop

_____ 2nd year's crop

_____ 3rd year's crop

KIND (X)

☐ Cattle

☐ Poultry

☐ Swine

☐ Other

TYPE (X)

Will manure be applied in liquid (slurry) form?

☐ Yes

☐ No

7. POTATOES ONLY

How severe is potato scab? (X)

☐ None

☐ Minor

☐ Moderate

☐ Severe

Will scab resistant varieties be grown on this field? (X)

☐ Yes

☐ No

8. ADDITIONAL SOIL TESTS

List soil tests desired in addition to **pH, P, K, Mg, Ca** and **lime requirement**
(See back of last page for list of additional soil tests available)

_____

_____

9. LAST CROP FERTILIZER USE

WHEN APPLIED | ANALYSES OR GRADE | RATE (Lbs /A)

Before planting _____ _____

At planting _____ _____

After planting or topdressing } _____ _____

Or list only TOTAL FERTILIZER USED | N | P_2O_5 | K_2O

10. CROP YIELD

ESTIMATE YIELD OF LAST CROP. CHECK (X) UNIT

YIELD

☐ Bu

☐ Boxes

☐ CWT

☐ Tons

11. PREVIOUS LIME USE

Years since lime was last applied (X)

☐ Last year

☐ 2 to 4 years

☐ 4 + years

Rate (Tons/Acre) of last application (X)

☐ < 1

☐ 1 to 2

☐ 2 to 4

☐ 4 +

12. COMMENTS

_____

_____

_____

13. TOTAL NO. OF SAMPLES SUBMITTED

TABLE 16/*Computer input form for the SCS Form 5 of soil survey interpretations.*

SCS-SOILS-5
REV. MAY 1972
FILE CODE SOILS-12

U.S. DEPARTMENT OF AGRICULTURE
SOIL CONSERVATION SERVICE

SOIL SURVEY INTERPRETATIONS

KEYING ONLY		
RECORD NO.	CONTROL	
	WORD	NO.
	MLRA	001
	STATE	011

MLRA(S) ☐   KIND OF UNIT ☐   UNIT NAME ☐

STATE ☐   RECORD NO. ☐   AUTHOR(S) ☐   DATE ☐   REVISED ☐   UNIT MODIFIER ☐

CLASSIFICATION AND BRIEF SOIL DESCRIPTION

	CLASS	021
	DESCR	031
		2
		3
		4
		5

ESTIMATED SOIL PROPERTIES

FOOTNOTE

		DEPTH (IN.)	USDA TEXTURE	UNIFIED	AASHO	FRACT. >3 IN. (PCT)	PERCENT OF MATERIAL LESS THAN 3 IN. PASSING SIEVE				LIQUID LIMIT	PLASTICITY INDEX
							4	10	40	200		
PROP	041											
	2											
	3											
	4											
	5											
	6											

		DEPTH (IN.)	PERMEABILITY (IN/HR)	AVAILABLE WATER CAPACITY (IN/IN)	SOIL REACTION (pH)	SALINITY (MMHOS/CM)	SHRINK-SWELL POTENTIAL	CORROSIVITY		EROSION FACTORS		WIND EROD. GROUP
								STEEL	CONCRETE	K	T	
PROP	051	SAME DEPTH AS ABOVE										
	2											
	3											
	4											
	5											
	6											

		FLOODING			HIGH WATER TABLE			CEMENTED PAN		BEDROCK		SUBSIDENCE		HYD GRP	POTENTIAL FROST ACTION
		FREQUENCY	DURATION	MONTHS	DEPTH (FT)	KIND	MONTHS	DEPTH (IN)	HARDNESS	DEPTH (IN)	HARDNESS	INITIAL (IN)	TOTAL (IN)		
PROP	061														

FOOTNOTES   **SANITARY FACILITIES**   KEYING ONLY   FOOTNOTES   **SOURCE MATERIAL**

SEPTIC	071	SEPTIC TANK ABSORPTION FIELDS	FILL	191	ROADFILL
	2			2	
	3			3	
	4			4	
	5			5	
LAGOON	081	SEWAGE LAGOONS	SAND	201	SAND
	2			2	
	3			3	
	4			4	
	5			5	
TRENCH	091	SANITARY LANDFILL (TRENCH)	GRAVEL	211	GRAVEL
	2			2	
	3			3	
	4			4	
	5			5	
SANARE	101	SANITARY LANDFILL (AREA)	SOIL	221	TOPSOIL
	2			2	
	3			3	
	4			4	
	5			5	
COVER	111	DAILY COVER FOR LANDFILL			

FOOTNOTES   **WATER MANAGEMENT**

	2		PONDRS	231	POND RESERVOIR AREA
	3			2	
	4			3	
	5			4	
				5	

FOOTNOTES   **COMMUNITY DEVELOPMENT**

EXCAV	121	SHALLOW EXCAVATIONS	DIKES	241	EMBANKMENTS DIKES AND LEVEES
	2			2	
	3			3	
	4			4	
	5			5	
DWEL	131	DWELLINGS WITHOUT BASEMENTS	PONDAQ	251	EXCAVATED PONDS AQUIFER FED
	2			2	
	3			3	
	4			4	
	5			5	
DWEL	141	DWELLINGS WITH BASEMENTS	DRAIN	261	DRAINAGE
	2			2	
	3			3	
	4			4	
	5			5	
BLDGS	151	SMALL COMMERCIAL BUILDINGS	IRRIG	271	IRRIGATION
	2			2	
	3			3	
	4			4	
	5			5	
ROADS	161	LOCAL ROADS AND STREETS	TERRAC	281	TERRACES AND DIVERSIONS
	2			2	
	3			3	
	4			4	
	5			5	

FOOTNOTES   **REGIONAL INTERPRETATIONS**

			WATERW	291	GRASSED WATERWAYS
REGION	171			2	
	2			3	
	3			4	
	4			5	
REGION	181				
	2				
	3				
	4				

TABLE 16 (continued)

Table 16 form (continued): a soil data coding form with the following sections and labels.

Header block:
- KEYING ONLY — RECORD NO., CONTROL (WORD, NO.)
- UNIT NAME: _____
- UNIT MODIFIER: _____
- (2)
- FOOTNOTE

Left column control words and numbers:
- CAMPS 301, 2, 3, 4, 5 — CAMP AREAS
- PICNIC 311, 2, 3, 4, 5 — PICNIC AREAS

RECREATION section (right):
- KEYING ONLY
- PLAYGD 321, 2, 3, 4, 5 — PLAYGROUNDS
- PATHS 331, 2, 3, 4, 5 — PATHS AND TRAILS
- FOOTNOTE

CAPABILITY AND PREDICTED YIELDS - CROPS AND PASTURE (HIGH LEVEL MANAGEMENT)

CLASS-DETERMINING PHASE	CAPABILITY																
	NIRR	IRR.	NIRR	IRR.	NIRR	IRR.	NIRR	IRR.	NIRR	IRR.	NIRR	IRR.	NIRR	IRR.	NIRR	IRR.

Control words: CROPHD 451, 2, 3; CROPS 341, 2, 3, 4, 5, 6, 7, 8, 9, 351, 2, 3

FOOTNOTE

WOODLAND SUITABILITY

CLASS-DETERMINING PHASE	ORD SYM	MANAGEMENT PROBLEMS					POTENTIAL PRODUCTIVITY		TREES TO PLANT
		EROSION HAZARD	EQUIP. LIMIT	SEEDLING MORT'Y.	WINDTH. HAZARD	PLANT COMPET.	IMPORTANT TREES	SITE INDEX	

Control words: WOODS 361, 2, 3, 4, 5, 6, 7, 8, 9, 371, 2, 3, 4, 5, 6

FOOTNOTE

WIND BREAKS

CLASS-DETERMINING PHASE	SPECIES	HT	SPECIES	HT	SPECIES	HT	SPECIES	HT

Control words: WINDBK 381, 2, 3, 4, 5, 6

FOOTNOTE

WILDLIFE HABITAT SUITABILITY

CLASS-DETERMINING PHASE	POTENTIAL FOR HABITAT ELEMENTS								POTENTIAL AS HABITAT FOR:			
	GRAIN & SEED	GRASS & LEGUME	WILD HERB.	HARDWD TREES	CONIFER PLANTS	SHRUBS	WETLAND PLANTS	SHALLOW WATER	OPENLAND WILDLIFE	WOODLAND WILDLIFE	WETLAND WILDLIFE	RANGELAND WILDLIFE

Control words: WILDLF 391, 2, 3, 4, 5, 6

FOOTNOTE

POTENTIAL NATIVE PLANT COMMUNITY (RANGELAND OR FOREST UNDERSTORY VEGETATION)

COMMON PLANT NAME	PLANT SYMBOL (NLSPN)	PERCENTAGE COMPOSITION (DRY WEIGHT) BY CLASS DETERMINING PHASE

Control words: PHASE 401, 2; PLANT 411, 2, 3, 4, 5, 6, 7, 8, 9, 421, 2, 3, 4, 5, 6

POTENTIAL PRODUCTION (LBS./AC. DRY WT):
- FAVORABLE YEARS
- NORMAL YEARS
- UNFAVORABLE YEARS

Control words: PRODUC 431, 2, 3

SYM. — FOOTNOTES

Control words: NOTES 441, 2, 3, 4, 5, 6, 7

TABLE 17/*Brief instructions for filling out the SCS Form 5 for input of soil interpretation information into the computer system.*

SOIL SURVEY INTERPRETATIONS INSTRUCTIONS

(Print clearly in CAPITAL letters. See instructions at bottom of page 2 for entering data legibly.)

1. MLRA(S)--List all the MLRA's (separated by commas) to which the full set of interpretations apply. If some interpretations do not apply to all MLRA's in which the soil occurs, prepare additional SCS-Soils-5's.
2. Kind of Unit--Enter the kind of unit: SERIES, VARIANT, LAND TYPE, GREAT GROUP, SUBGROUP, FAMILY or FAMILY PHASE. If great group, subgroup, family or family phase is entered, the full classification name should be printed on the first line of the block "Classification and Brief Soil Description".
3. Unit Name--Print name of series, variant, land type, etc. (Use no more than 29 characters including spaces). For CNI, if the soil name on diagnostic is in code, print the code in "Unit Name" space.
4. State--Print full name of the state having responsibility for the series.
5. Record No.--State assign a unique sequential number to each Soils-5 prepared, (1 for the first Soils-5 prepared, 2 for the second, etc.) This number is used by the computer to identify each Soils-5. If a Soils-5 is revised, the original record number must be retained.
6. Author(s), Date, Revised--Enter author(s) initials and the date on which the Soils-5 was first prepared or revised. If revised, enter an X in the space provided.
7. Unit Modifier--Enter the soil property which is used as phase criterion that so drastically changes any interpretation that a separate Soils-5 must be prepared, e.g., STONY, SALINE or GRAVELLY SUBSTRATUM.
8. Classification and Brief Soil Description--On the first line (Class 021), print the full classification name of a variant, great group, subgroup, family or family phase. Do not enter classification of series or land types (the computer will print the classification of the series from the current classification file). On lines DESCR 031-035, either print or type (with no more than 120 characters per line and 600 characters total) a short narrative description in non-technical language. This description should not conflict with the standard series description. (1) In the lead sentence give one or two of the major features that characterize the soil and are important to their use and behavior, such as, depth and drainage class. (2) Describe the setting for the series including position on the landscape, shape and ranges of slope names and the parent material where known with reasonable confidence. (3) Describe layers abstracted from the "Estimated Soil Properties" block. Color, texture and thickness of the layers are commonly covered along with special features if they are important to use and management. (4) Give the full range of slope of the series in percent. (See p. 19-20, "Guide to Authors of Manuscripts for Published Soil Surveys".)

ESTIMATED SOIL PROPERTIES

(See "Guide for Interpreting Engineering Uses of Soil" for explanation of many of the following items.)

1. Footnote--Enter a capital letter, e.g., A, if a footnote applies to the whole "Estimated Soil Properties" block. See instructions on page 2 under "footnotes".
2. Depth--A maximum of six layers can be accomodated and up to three sets of surface texture that differ in estimated properties.
3. USDA Texture--Up to three textures can be entered on each line. Separate them by commas. If modifiers are used, they must be attached to the texture by a hyphen, e.g., GR-SL. If a layer is stratified, enter SR as a modifier and the end members of the textural range all connected by hyphens, e.g., SR-S-L.

Modifier:		Texture or terms used in lieu of texture:			
BY	Bouldery	COS	Coarse sand	CE	Coprogenous earth
BYV	Very bouldery	S	Sand	DE	Diatomaceous earth
BYX	Extremely bouldery	FS	Fine sand	FB	Fibric material
CB	Cobbly	VFS	Very fine sand	HM	Hemic material
CBA	Angular cobbly	LCOS	Loamy coarse sand	ICE	Ice or frozen soil
CBV	Very cobbly	LS	Loamy sand	IND	Indurated
CN	Channery	LFS	Loamy fine sand	MARL	Marl
CR	Cherty	LVFS	Loamy very fine sand	MPT	Mucky-peat
CRC	Coarse cherty	COSL	Coarse sandy loam	MUCK	Muck
CRV	Very cherty	SL	Sandy loam	PEAT	Peat
FL	Flaggy	FSL	Fine sandy loam	SP	Sapric material
GR	Gravelly	VFSL	Very fine sandy loam	UWB	Unweathered bedrock
GRC	Coarse gravelly	L	Loam	VAR	Variable
GRF	Fine gravelly	SIL	Silt loam	WB	Weathered bedrock
GRV	Very gravelly	SI	Silt		
MK	Mucky	SCL	Sandy clay loam		
PT	Peaty	CL	Clay loam		
SH	Shaly	SICL	Silty clay loam		
SHV	Very shaly	SC	Sandy clay		
SR	Stratified	SIC	Silty clay		
ST	Stony	C	Clay		
STV	Very stony				
STX	Extremely stony				
SY	Slaty				

4. Unified--Enter up to 4 classes. Separate by commas, e.g., CL-ML, ML. Enter no dual classes except CL-ML.
5. AASHO--Enter up to 4 classes. Separate by commas, e.g., A-6-7, A-7.
6. Fraction$>$3 in. (Pct)--Enter the weight percentage of material greater than 3 inches, e.g., 30-60. Enter a zero (0) if none occurs.
7. Percent of Material Less Than 3 inches Passing Sieve--Enter the range in weight percentages passing each of the sieve sizes, e.g., 80-100.
8. Liquid Limit--Enter the range of liquid limit, e.g., 20-30, or NP if the soil is nonplastic.
9. Plasticity Index--Enter the range of plasticity index, e.g., 10-20, or NP if the soil is nonplastic.
10. Permeability--Use the following classes: $<$0.06, 0.06-0.2, 0.2-0.6, 0.6-2.0, 2.0-6.0, 6.0-20, $>$20. Classes may be combined, e.g., 0.06-0.6.
11. Available Water Capacity--Enter the estimated range of available water capacity in inches per inch, e.g., .10-.15.
12. Soil Reaction (pH)--Enter the range of pH (1:1 water). Commonly the class ranges are given or combination of classes. The classes are 3.5, 3.6-4.4, 4.5-5.0, 5.1-5.5, 5.6-6.0, 6.1-6.5, 6.6-7.3, 7.4-7.8, 7.9-8.4, 8.5-9.0, 9.0.
13. Salinity (mmhos/cm)--Give a range of the electrical conductivity of the saturation extract during the growing season, e.g., 2-10. If salinity is no problem for growing plants, enter a dash.
14. Shrink-Swell Potential--Use one of the following classes: LOW, MODERATE, or HIGH.
15. Corrosivity--Classes are: LOW, MODERATE or HIGH.
16. Erosion Factors--List coordinated K and T factors. List K factors for each major horizon if they are significantly different; T factors for surface layer(s) only. On soils with less than 2 or 3 percent slope, enter a dash.
17. Wind Erodibility Group--Enter wind erodibility group for surface layer(s) only. In parts of the country where wind erosion is no problem, enter a dash.
18. Flooding--Define the natural, unprotected soil in terms of frequency. Duration and months that floods are likely to occur are given only for soils that flood more frequently than rare. Ranges of frequency and duration classes may be given if needed, e.g., RARE-COMMON, BRIEF-V. LONG.

    Frequency: NONE (No reasonable possibility of flooding)
    RARE (Flooding unlikely but possible under abnormal conditions)
    COMMON (Flooding likely under normal conditions)
    OCCASIONAL (Less often than once in 2 years)
    FREQUENT (More often than once in 2 years)

    Duration: V. BRIEF (Less than 2 days)
    BRIEF (2 days to 7 days)
    LONG (7 days to 1 month)
    V. LONG (More than 1 month)

    Months: Give months of probable flooding, use abbreviations, e.g., NOV-MAR.

19. High Water Table--Give the depth range of seasonally high water table to the nearest half-foot, e.g., 1.5-3. Enter APPARENT under kind of water table unless known that the water table is perched, then enter PERCHED. Enter the months in which water table is likely to be within the normal depth of observation, e.g., NOV-MAR. If the water table is below 6 feet or if water table exists for less than one month, enter $>$6.
20. Cemented Pan--Enter depth range in inches to a cemented pan such as a duripan, petrocalcic, petrogypsic, or ortstein layer. Enter RIPPABLE or HARD in the hardness column. (Rippable and hard are defined under "bedrock" below). If soil has no cemented pan, enter a dash.
21. Bedrock--Enter depth range in inches to bedrock and hardness of rock. Enter RIPPABLE or HARD in the hardness column. "Rippable" rock can be excavated using a single tooth ripping attachment mounted on a 200-300 horsepower tractor. "Hard" rock requires blasting or use of excavators larger than 200-300 horsepower (see Section 2 of Specialification 21, National Engineering Handbook for additional equipment requirments). If depth to bedrock is below the normal depth of observation enter $>$60.
22. Subsidence--Give depth range in inches of initial and total drainage induced subsidence of organic soils or other wet soils that subside when drained. If subsidence is not a problem, enter a dash.
23. Hydrologic Group--Give the coordinated hydrologic group letter (A,B,C,D,A/D,B/D, or C/D.
24. Potential Frost Action--Enter one of the following: LOW, MODERATE or HIGH. In parts of the country where frost action is no problem, enter a dash.

INTERPRETATIONS FOR SELECTED USES

Give soil limitation or suitability ratings for the selected uses. (Refer to "Guide for Interpreting Engineering Uses of Soil".) Items affecting use that are used as phase criterion are entered in the block first, followed by the limitation or suitability ratings and then the restrictions that apply, e.g., 2-7%: MODERATE-SLOPE. Items used as phase criteria are slope, flooding, texture of surface layer, stoniness and depth to rock. For sloping soils, enter the slope breaks that cause different ratings for each use, e.g., for septic tank absorption fields, the slope breaks are 0-8, 8-15, 15+, for sewage lagoons the slope breaks are 0-2, 2-7, 7+. If flood prone soils are given adequate protection for a given use, rate both the "flooded phase" and the "protected phase", e.g., for dwellings--COMMON: SEVERE-FLOODS and PROTECTED: SLIGHT. Punctuation is important--use colon between phase designation and rating; hyphen between limitations and restrictions; commas to separate any series of items and use no periods. Enter slope first if it is used with additional items affecting use, e.g., 8-15% SCL, CL: MODERATE-SLOPE, TOO CLAYEY. In the following blocks are the rating terms and restrictive features to be used for each use.

SANITARY FACILITIES

Use	Rating	Restrictive Features			
Septic Tank	SLIGHT	CEMENTED PAN	LARGE STONES	SLOPE	
Absorption	MODERATE	DEPTH TO ROCK	PERCS SLOWLY	WET	
Fields	SEVERE	FLOODS	ROCK OUTCROPS		
	SLIGHT	CEMENTED PAN	FLOODS	SLOPE	
Sewage	MODERATE	DEPTH TO ROCK	LARGE STONES	SMALL STONES	
Lagoons	SEVERE	EXCESS HUMUS	PERCS SLOWLY	WET	
Sanitary	SLIGHT	CEMENTED PAN	FLOODS	ROCK OUTCROPS	TOO SANDY
Landfill	MODERATE	DEPTH TO ROCK	LARGE STONES	SLOPE	WET
(Trench)	SEVERE	EXCESS HUMUS	PERCS SLOWLY	TOO CLAYEY	
Sanitary	SLIGHT	FLOODS	WET		
Landfill	MODERATE	PERCS RAPIDLY			
(Area)	SEVERE	SLOPE			
Daily	GOOD	EXCESS HUMUS	PERCS RAPIDLY	THIN LAYER	WET
Cover for	FAIR	HARD TO PACK	SLOPE	TOO CLAYEY	
Landfill	POOR	LARGE STONES	SMALL STONES	TOO SANDY	

COMMUNITY DEVELOPMENT

Use	Rating	Restrictive Features			
Shallow	SLIGHT	CEMENTED PAN	EXCESS HUMUS	ROCK OUTCROPS	TOO CLAYEY
Excavations	MODERATE	CUTBANKS CAVE	FLOODS	SLOPE	WET
	SEVERE	DEPTH TO ROCK	LARGE STONES	SMALL STONES	
Dwellings	SLIGHT	CEMENTED PAN	FLOODS	LOW STRENGTH	SLOPE
Without	MODERATE	DEPTH TO ROCK	FROST ACTION	ROCK OUTCROPS	WET
Basements	SEVERE	EXCESS HUMUS	LARGE STONES	SHRINK-SWELL	
Dwellings	SLIGHT	CEMENTED PAN	FLOODS	LOW STRENGTH	SLOPE
With	MODERATE	DEPTH TO ROCK	FROST ACTION	ROCK OUTCROPS	WET
Basements	SEVERE	EXCESS HUMUS	LARGE STONES	SHRINK-SWELL	
Small	SLIGHT	CORROSIVE	FLOODS	LOW STRENGTH	SLOPE
Commercial	MODERATE	DEPTH TO ROCK	FROST ACTION	ROCK OUTCROPS	WET
Buildings	SEVERE	EXCESS HUMUS	LARGE STONES	SHRINK-SWELL	
Local	SLIGHT	CEMENTED PAN	FLOODS	LOW STRENGTH	SLOPE
Roads and	MODERATE	DEPTH TO ROCK	FROST ACTION	ROCK OUTCROPS	WET
Streets	SEVERE	EXCESS HUMUS	LARGE STONES	SHRINK-SWELL	

REGIONAL INTERPRETATIONS

Interpretations approved for use within the region may be added in these blocks.

SOURCE MATERIALS

Use	Rating	Restrictive Features			
	GOOD	AREA RECLAIM	LARGE STONES	SHRINK-SWELL	WET
Roadfill	FAIR	EXCESS HUMUS	LOW STRENGTH	SLOPE	
	POOR	FROST ACTION	ROCK OUTCROPS	THIN LAYER	
	GOOD				
Sand	FAIR				
	POOR				
	UNSUITED				
	GOOD				
Gravel	FAIR				
	POOR				
	UNSUITED				
	GOOD	AREA RECLAIM	EXCESS SALT	SMALL STONES	TOO SANDY
Topsoil	FAIR	EXCESS ALKALI	LARGE STONES	THIN LAYER	WET
	POOR	EXCESS LIME	SLOPE	TOO CLAYEY	

WATER MANAGEMENT

Use	Rating 1/	Features Affecting			
Pond	SLIGHT 1/	CEMENTED PAN	PERCS RAPIDLY		
Reservoir	MODERATE 1/	DEPTH TO ROCK	SLOPE		
Area	SEVERE 1/	FAVORABLE			
	SLIGHT 1/	COMPRESSIBLE	LARGE STONES	PIPING	UNSTABLE FILL
Pond	MODERATE 1/	FAVORABLE	LOW STRENGTH	ROCK OUTCROPS	
Embankment	SEVERE 1/	HARD TO PACK	PERCS RAPIDLY	THIN LAYER	
Excavated	SLIGHT 1/	DEEP TO WATER	NO WATER		
Ponds	MODERATE 1/	FAVORABLE	ROCK OUTCROPS		
Aquifer Fed	SEVERE 1/	LARGE STONES	SLOW REFILL		
		CEMENTED PAN	DEPTH TO ROCK	FAVORABLE	PERCS SLOWLY
Drainage		COMPLEX SLOPE	EXCESS ALKALI	FLOODS	POOR OUTLETS
		CUTBANKS CAVE	EXCESS SALT	NOT NEEDED	SLOPE
					WET
		COMPLEX SLOPE	EXCESS ALKALI	FAST INTAKE	ROOTING DEPTH
Irrigation		DROUGHTY	EXCESS LIME	FAVORABLE	SLOW INTAKE
		ERODES EASILY	EXCESS SALT	PERCS RAPIDLY	WET
Terraces		CEMENTED PAN	ERODES EASILY	NOT NEEDED	ROCK OUTCROPS
And		COMPLEX SLOPE	FAVORABLE	PERCS SLOWLY	ROOTING DEPTH
Diversions		DEPTH TO ROCK	LARGE STONES	PIPING	SLOPE
				POOR OUTLETS	WET
		DROUGHTY	EXCESS SALT	NOT NEEDED	ROCK OUTCROPS
Grassed		ERODES EASILY	FAVORABLE	PERCS SLOWLY	SLOPE
Waterway		EXCESS ALKALI	LARGE STONES	ROOTING DEPTH	WET

1/ Limitation ratings and restrictions may be used instead of "features affecting" where regional criteria have been developed.

TABLE 17 *(continued)*

SOIL SURVEY INTERPRETATIONS INSTRUCTIONS

INTERPRETATIONS FOR SELECTED USES (CONTINUED)

RECREATION

Enter the data following the instructions under "Interpretations for Selected Uses". Use Soils Memorandum-69 as a guide in making the ratings. In the following blocks are the rating terms and restrictions features to be used for each use.

Use	Rating		Restrictive Features			
Camp Areas	SLIGHT	DUSTY	PERCS SLOWLY	SMALL STONES	WET	
	MODERATE	FLOODS	ROCK OUTCROPS	TOO CLAYEY		
	SEVERE	LARGE STONES	SLOPE	TOO SANDY		
Picnic Areas	SLIGHT	DUSTY	ROCK OUTCROPS	TOO CLAYEY		
	MODERATE	FLOODS	SLOPE	TOO SANDY		
	SEVERE	LARGE STONES	SMALL STONES	WET		

Use	Rating		Restrictive Features			
Play Grounds	SLIGHT	DEPTH TO ROCK	LARGE STONES	SLOPE	TOO SANDY	
	MODERATE	DUSTY	PERCS SLOWLY	SMALL STONES	WET	
	SEVERE	FLOODS	ROCK OUTCROPS	TOO CLAYEY		
Paths And Trails	SLIGHT	DUSTY	ROCK OUTCROPS	TOO CLAYEY		
	MODERATE	FLOODS	SLOPE	TOO SANDY		
	SEVERE	LARGE STONES	SMALL STONES	WET		

CAPABILITY AND PREDICTED YIELDS - CROPS AND PASTURE (HIGH LEVEL MANAGEMENT)

1. Class-Determining Phase--For each phase that significantly influences yield or management, give the coordinated capability classification and predicted yields of major cultivated crops, hay and pasture commonly grown on the soil. Phases that commonly influence yield and management significantly are flooding, drainage, slope, texture of surface layer and erosion. Enter OCCASIONAL or FREQUENT if flooding influences yield or management. If flood prone soils are given adequate protection for crops, enter PROTECTED in the class-determining phase column and give capability class and subclass and yields. Give capability class and subclass and yields for DRAINED soil if drainage is feasible or UNDRAINED soil if drainage is not feasible. If both drained and undrained phases occur, capability and yields should be given for each. On sloping soils, use the slope groups that are most common in the MLRA in which the typifying pedon is located. Enter SEV.ER. for severely eroded phases that influence yields and management (texture of surface is likely to take on the character of the sub-soil). If moderately eroded phases influence yields and management, enter ERODED. If more than one phase is given, list slope first, texture second, erosion third, e.g., 5-8% CL, SEV.ER. If all phases are rated alike, write ALL in this column.

2. Capability--Give the nonirrigated capability class and subclass for all soils in the column headed NIRR. If the soil is irrigated or is potential irrigated land, enter the irrigated capability class and subclass in the IRR column. Use arabic numbers, 3W, not IIIW.

3. Predicted Yields--On lines CROPHD 451, and 452 if needed, enter the name of the crop. On line CROPHD 453 enter the unit of measure, e.g., (BU), (TONS), (CWT), etc. For each class-determining phase, give the predicted yield of crops approximating those obtained by leading commercial farmers at the level of management which tends to produce the highest economic returns per acre (commonly known as "B" level management). "B" level management includes using the best varieties; balancing plant populations and added plant nutrients to the potential of the soil; control of erosion, weeds, insects and diseases; maintenance of optimum soil tilth; adequate soil drainage; and timely operations. List the yields of the common crops grown in the area in the NIRR (nonirrigated) column, the IRR (irrigated) column, or both columns. Examples of crops and units of measure that are commonly used are as follows:

BARLEY	(BU)	LEGUME HAY	(TONS)	SOYBEANS	(BU)
CORN	(BU)	OATS	(BU)	SUGAR BEETS	(TONS)
CORN SILAGE	(TONS)	PASTURE	(AUM)	SUGAR CANE	(TONS)
COTTON	(LBS LINT)	PEANUTS	(LBS)	TOBACCO	(LBS)
GRAIN SORGHUM	(BU)	POTATOES	(CWT)	TOMATOES	(TONS)
GRASS HAY	(TONS)	RICE	(BU)	WHEAT	(BU)
GRASS-LEGUME HAY	(TONS)	SORGHUM SILAGE	(TONS)		

WOODLAND SUITABILITY

1. If woodland is not an important segment of land use, enter NONE on the first line in the column "Important Trees".

2. Rate only those phases that determine the potential productivity or that are class determining in terms of management. If all phases of a series are rated alike write ALL in the column headed "Class-determining phase".

3. Give only the first two elements of the ordination symbol--the class and the subclass. The third element (group) is regional or local in nature and cannot be coordinated nationally at this time.

4. Follow Soils Memorandum-26, Rev. 2 for guidance in making the ratings, SLIGHT, MODERATE, or SEVERE in the management problem columns.

5. List several of the indicator tree species or forest types and the site index of each in the column "Important Trees". If the site index for a tree species is a summary of 5 or more actual measurements on this soil, enter an asterisk (*) after the index number. A footnote will automatically be printed by the computer as follows: "*Site index is a summary of 5 or more measurements on this soil." Do not enter this footnote on form Soils-5.

6. List one or more tree species suitable for planting. In areas where important, Christmas tree species may be shown in the column "Trees to Plant" followed by a double asterisk, e.g., Arizona Cypress**. A footnote will automatically be printed by the computer as follows: "**Christmas tree species." Do not enter this footnote on form Soils-5. If additional footnotes are needed in this column, they may be entered following the tree species, e.g., SLASH PINE 5/.

WINDBREAKS

Enter the important windbreak tree species and expected height at 20 years for the class-determining phases. If all phases grow the same species, write ALL in the "Class-Determining phase" column. Soils Memorandum-64 contains information on obtaining data for shelterbelts and windbreaks. In parts of the country where windbreaks are not normally needed, enter NONE on the first line under "Species".

WILDLIFE HABITAT SUITABILITY

1. Rate only those phases that are classes determining for potential habitat elements. If all phases of a series are rated alike, write ALL in the "Class-Determining Phase" column. Rating terms are GOOD, FAIR, POOR, V. POOR (to stand for very poor). If an element is not rated, enter a dash.

2. Give a summary rating for the kinds of wildlife. Soils rated for woodland wildlife generally will not be rated for rangeland wildlife and vice versa. If a kind of wildlife is not rated, enter a dash.

3. Use Soils Memorandum-74 as a guide in making ratings.

POTENTIAL NATIVE PLANT COMMUNITY (RANGELAND OR FOREST UNDERSTORY VEGETATION)

1. Enter the names of the class-determining phases for each percentage composition column on line PHASEH 401 and, if additional space is needed, line PHASEH 402, e.g., LOAMY SAND in the first percentage composition column and SANDY LOAM in the second.

2. List the common name of the major native plants that grow under climax condition on this soil and show the plant symbol from the National List of Scientific Plant Names, USDA, SCS, 1971. In rangeland, the species will be grasses, shrubs and forbs and in Savannah sites, trees also are included. Indicate by footnote those rangeland plants that are not usually utilized by cattle or sheep, e.g., YUCCA 6/. In woodland, understory species are listed for grasses, shrubs, forbs and other understory plants within reach of livestock or grazing or browsing wildlife. Understory species and composition should be based on the canopy density which most nearly represents the highest wood production for the forest plant community.

3. For each of the class-determining phases show the percent composition (dry weight) for the major species. Enter OTHER in the plant symbol column and give percentage composition of minor species so that the total is 100 percent.

4. Enter total potential production for favorable years, normal years and unfavorable years.

5. Where data are not available and acceptable estimates cannot be made, list the species in order of their general productivity and leave columns for percent composition blank.

6. Refer to the National Handbook for Range and Related Grazing Lands for additional instructions.

FOOTNOTES

1. Footnote symbols are entered in the "footnote" columns provided in each interpretation block. Enter a capital letter for footnotes that refer to major headings such as Estimated Soil Properties, Sanitary Facilities, etc., and numerals for subordinate headings such as septic tank filter fields, sewage lagoons, etc. The corresponding footnotes should be printed out in the space at the bottom of the second page.

2. No more than one line may be used for any one footnote. For example, to footnote the major heading "Sanitary Facilities" enter a letter A in the footnote column (SEPTIC 071) and on line NOTES 441, enter A in the SYM (symbol) column and print the footnote, RATINGS BASED ON "GUIDE FOR INTERPRETATIONS ENGINEERING USES OF SOILS", NOV. 1971. Footnote the "Septic Tank Absorption Field" heading by entering a number 1 in the footnote column (line SEPT 072), entering a number 1 in the SYM column of line NOTES 442 and print the footnote, EXCESSIVE PERMEABILITY MAY CAUSE POLLUTION OF GROUND WATER. The following are examples of some of the more common footnotes:

ESTIMATES BASED ON ENGINEERING TEST DATA OF (number) PEDONS FROM (counties, states).
RATINGS BASED ON "GUIDE FOR INTERPRETING ENGINEERING USES OF SOILS", NOV. 1971.
EXCESSIVE PERMEABILITY RATE MAY CAUSE POLLUTION OF GROUND WATER.
RECREATION RATINGS BASES ON SOILS MEMORANDUM-69, OCT. 1968.
WILDLIFE RATING BASED ON SOILS MEMORANDUM-74, JAN. 1972.
NOT USUALLY UTILIZED BY CATTLE OR SHEEP.

INSTRUCTIONS FOR ENTERING DATA LEGIBLY FOR KEYPUNCHING EFFICIENCY

1. Make no entries where data are not available or reasonable estimates cannot be made.

2. Do not enter data in "Keying Only" columns.

3. Make entries in black pencil with hardness rating of 2½, H or HB. Entries may be typed, however, take care not to overfill the space (This form has 10 characters per inch, elite type has 12).

4. Print clearly in CAPITAL letters. The following pairs of letters are often confused. Be careful to print them legibly: AH, LC, UV. Certain letters and numbers are also confused: OO, I1, Z2, S5. Where possibility of confusion exists, use the following symbols for the letters: Ø, I, Z. Make sure the S has round curves and the 5 has sharp corners.

5. Punctuation is important. Follow the instructions and examples give for each of the entries. When using hyphens and decimal points make sure each can be identified.

6. If a block has more than one line, always make entry on first line, use each in order. DO NOT SKIP LINES.

7. If a line data entry in a column is the same below, an arrow can be used to indicate the entry is duplicated, e.g.,

SHRINK-SWELL POTENTIAL
LOW
↓

8. Be sure to PROOF READ.

TABLE 18/*Computer output form for the SCS Form 5 of soil survey interpretations.*

```
CT0011                    S O I L   S U R V E Y   I N T E R P R E T A T I C N S

MLRA(S): 142, 144                                                              STOCKBRIDGE SERIES
REV. WNG-EHS, 1-75
DYSTRIC ELTROCHREPTS, COARSE-LOAMY, MIXED, MESIC
```

THE STOCKBRIDGE SERIES CONSISTS OF DEEP, WELL DRAINED SOILS ON UPLANDS. THEY FORMED IN GLACIAL TILL. TYPICALLY THESE SOILS HAVE A DARK BROWN LOAM SURFACE LAYER 10 INCHES THICK. THE SUBSOIL LAYERS FROM 10 TO 28 INCHES ARE OLIVE BROWN AND DARK GRAYISH BROWN LOAM. THE SUBSTRATUM FROM 28 TO 54 INCHES IS OLIVE LOAM WITH SOME FRAGMENTS OF LIMESTONE. SLOPES RANGE FROM 0 TO 45 PERCENT.

ESTIMATED SOIL PROPERTIES (A)

DEPTH (INₐ)	USCA TEXTURE	UNIFIED	AASHTO	FRACT >3 IN (PCT)	PERCENT OF MATERIAL LESS THAN 3" PASSING SIEVE NO. 4	10	40	200	LIQUID LIMIT	PLAS- TICITY INDEX
0-10	GR-FSL, GR-L	ML, CL-ML	A-4	0-5	60-80	55-75	40-65	20-40	25-40	4-8
0-10	L, SIL, GR-SIL	ML, CL-ML	A-4	0-5	75-95	70-90	65-85	55-75	25-40	4-8
10-28	L, SIL, GR-SIL	ML, CL-ML	A-4	0-5	75-95	70-90	60-85	55-75	20-40	3-8
28-54	L, SIL, GR-L	ML, CL-ML	A-4	0-5	70-95	70-90	60-80	55-75	<40	NP-8

DEPTH (INₐ)	PERMEABILITY (IN/HR)	AVAILABLE WATER CAPACITY (IN/IN)	SOIL REACTION (PH)	SALINITY (MMHOS/CM)	SHRINK- SWELL POTENTIAL	CORROSIVITY STEEL	CONCRETE	EROSION FACTORS K	T	WIND EROD. GROUP
0-10	0.6-2.0	0.08-0.13	4.5-6.5	–	LOW	MODERATE	LOW	.28	3	–
0-10	0.6-2.0	0.11-0.28	4.5-6.5	–	LOW	MODERATE	LOW	.28	3	–
10-28	0.6-2.0	0.08-0.24	4.5-6.5	–	LOW	MODERATE	LOW	.43		
28-54	0.06-0.2	0.10-0.14	5.6-7.3	–	LOW	MODERATE	LOW	.17		

FLOODING			HIGH WATER TABLE			CEMENTED PAN		BEDROCK		SUBSIDENCE		HYD GRP	POTENT'L FROST ACTION
FREQUENCY	DURATION	MONTHS	DEPTH (FT)	KIND	MONTHS	DEPTH (IN)	HARDNESS	DEPTH (IN)	HARDNESS	INIT. (IN)	TOTAL (IN)		
NONE			3.0-6.0	PERCHED	DEC-APR	–		>60		–		B	MODERATE

SANITARY FACILITIES (B)

				SOURCE MATERIAL (B)	
SEPTIC TANK ABSORPTION FIELDS	0-15%: SEVERE-PERCS SLOWLY 15+%: SEVERE-SLOPE,PERCS SLOWLY		ROADFILL	0-15%: FAIR-FROST ACTION 15-25%: FAIR-SLOPE 25+%: POOR-SLOPE	
SEWAGE LAGOON AREAS	0-2%: MODERATE-SEEPAGE,WETNESS 2-7%: MODERATE-SLOPE,SEEPAGE,WETNESS 7+%: SEVERE-SLOPE		SAND	UNSUITED-EXCESS FINES	
SANITARY LANDFILL (TRENCH)	0-25%: SEVERE-WETNESS 25+%: SEVERE-SLOPE,WETNESS		GRAVEL	POOR-EXCESS FINES	
SANITARY LANDFILL (AREA)	0-8%: MODERATE-WETNESS 8-15%: MODERATE-SLOPE,WETNESS 15+%: SEVERE-SLOPE		TOPSOIL	0-8% L,SIL: FAIR-SMALL STONES 8-15% L,SIL: FAIR-SLOPE,SMALL STONES 15+% L,SIL: POOR-SLOPE 0-15% GR: POOR-SMALL STONES 15+% GR: POOR-SLOPE,SMALL STONES	
DAILY COVER FOR LANDFILL	0-8%: GOOD 8-15%: FAIR-SLOPE 15+%: POOR-SLOPE				

WATER MANAGEMENT (B)

			POND RESERVOIR AREA	SEEPAGE,SLOPE	

COMMUNITY DEVELOPMENT (B)

SHALLOW EXCAVATIONS	0-8%: MODERATE-WETNESS 8-15%: MODERATE-SLOPE,WETNESS 15+%: SEVERE-SLOPE		EMBANKMENTS DIKES AND LEVEES	SEEPAGE	
DWELLINGS WITHOUT BASEMENTS	0-8%: SLIGHT 8-15%: MODERATE-SLOPE 15+%: SEVERE-SLOPE		EXCAVATED PONDS AQUIFER FED	NO WATER	
DWELLINGS WITH BASEMENTS	0-8%: MODERATE-WETNESS 8-15%: MODERATE-SLOPE,WETNESS 15+%: SEVERE-SLOPE		DRAINAGE	NOT NEEDED	
SMALL COMMERCIAL BUILDINGS	0-4%: SLIGHT 4-8%: MODERATE-SLOPE 8+%: SEVERE-SLOPE		IRRIGATION	SLOPE	
LOCAL ROADS AND STREETS	0-8%: SLIGHT 8-15%: MODERATE-SLOPE 15+%: SEVERE-SLOPE		TERRACES AND DIVERSIONS	SLOPE	

REGIONAL INTERPRETATIONS (C)

LAWNS, LANDSCAPING, AND GOLF FAIRWAYS	0-8%: SLIGHT 8-15%: MODERATE-SLOPE 15+%: SEVERE-SLOPE		GRASSED WATERWAYS	SLOPE	

TABLE 18 *(continued)*

STOCKBRIDGE SERIES   CT0011

```
------------------------------------------- RECREATION (D) -------------------------------------------
|            | 0-8% L.SIL: SLIGHT                    ||             | 0-2% L.SIL: MODERATE-SMALL STONES       |
|            | 8-15% L.SIL: MODERATE-SLOPE           ||             | 2-6% L.SIL: MODERATE-SLOPE,SMALL STONES |
| CAMP AREAS | 0-8% GR: MODERATE-SMALL STONES        || PLAYGROUNDS | 6+% L.SIL: SEVERE-SLOPE                  |
|            | 8-15% GR: MODERATE-SLOPE,SMALL STONES ||             | 0-6% GR: SEVERE-SMALL STONES            |
|            | 15+%: SEVERE-SLOPE                    ||             | 6+% GR: SEVERE-SLOPE,SMALL STONES       |
|            | 0-8% L.SIL: SLIGHT                    ||             | 0-15%: SLIGHT                           |
|            | 8-15% L.SIL: MODERATE-SLOPE           || PATHS       | 15-25%: MODERATE-SLOPE                   |
|PICNIC AREAS| 0-8% GR: MODERATE-SMALL STONES        || AND         | 0-15% GR: MODERATE-SMALL STONES         |
|            | 8-15% GR: MODERATE-SLOPE,SMALL STONES || TRAILS      | 15-25% GR: MODERATE-SLOPE,SMALL STONES  |
|            | 15+%: SEVERE-SLOPE                    ||             | 25+%: SEVERE-SLOPE                      |
```

--------- CAPABILITY AND PREDICTED YIELDS -- CROPS AND PASTURE (HIGH LEVEL MANAGEMENT) ---------

CLASS-DETERMINING PHASE	CAPABILITY SYM	CORN SILAGE (TONS)		ALFALFA HAY (TONS)		GRASS-LEGUME HAY (TONS)		GRASS HAY (TONS)		PASTURE (AUM)						
		NIRR	IRR.	NIRR	IRR.	NIRR	IRR.	NIRR	IRR.	NIRR	IRR.	NIRR	IRR.	NIRR	IRR.
0-3%	1	24		5.0		4.5		4.5		9.5					
3-8%	2E	24		5.0		4.5		4.5		9.5					
8-15%	3E	22		5.0		4.5		4.5		9.5					
8-15% SEV. ER.	4E	20		4.5		4.0		4.0		9.5					
15-25%	4E	20		4.5		4.0		4.0		8.5					
15-25% SEV. ER.	6E	-		-		-		-		-					
25-35%	6E	-		-		-		-		-					
25-35% SEV. ER.	7E	-		-		-		-		-					
25+%	7E														

------------------------------------ WOODLAND SUITABILITY (D) ------------------------------------

CLASS-DETERMINING PHASE	ORD SYM	MANAGEMENT PROBLEMS					POTENTIAL PRODUCTIVITY		TREES TO PLANT
		EROSION HAZARD	EQUIP. LIMIT	SEEDLING MORT'Y.	WINDTH. HAZARD	PLANT COMPET.	IMPORTANT TREES	SITE INDX	
0-15%	3O	SLIGHT	SLIGHT	SLIGHT	SLIGHT		NORTHERN RED OAK	70	EASTERN WHITE PINE
15-35%	3R	SLIGHT	MODERATE	SLIGHT	SLIGHT		SUGAR MAPLE	6O	WHITE SPRUCE
35+%	3R	MODERATE	SEVERE	SLIGHT	SLIGHT		EASTERN WHITE PINE	75	NORWAY SPRUCE
									EUROPEAN LARCH

--------------------------------------- WINDBREAKS ---------------------------------------

CLASS-DETERMIN'G PHASE	SPECIES	HT	SPECIES	HT	SPECIES	HT	SPECIES	HT
	NONE							

------------------------------- WILDLIFE HABITAT SUITABILITY (D) -------------------------------

CLASS-DETERMINING PHASE	POTENTIAL FOR HABITAT ELEMENTS								POTENTIAL AS HABITAT FOR:				
	GRAIN & SEED	GRASS & LEGUME	WILD HERB.	HARDWD TREES	CONIFER PLANTS	SHRUBS	WETLAND PLANTS	SHALLOW WATER	OPENLD WILDLF	WOODLD WILDLF	WETLAND WILDLF	RANGELD WILDLF
0-3%	GOOD	GOOD	GOOD	GOOD	GOOD	-	POOR	V. POOR	GOOD	GOOD	V. POOR	-
3-8%	FAIR	GOOD	GOOD	GOOD	GOOD	-	POOR	V. POOR	GOOD	GOOD	V. POOR	-
8-15%	FAIR	GOOD	GOOD	GOOD	GOOD	-	V. POOR	V. POOR	GOOD	GOOD	V. POOR	-
15-25%	POOR	FAIR	GOOD	GOOD	GOOD	-	V. POOR	V. POOR	FAIR	GOOD	V. POOR	-
25-35%	V. POOR	FAIR	GOOD	GOOD	GOOD	-	V. POOR	V. POOR	FAIR	GOOD	V. POOR	-
35+%	V. POOR	POOR	GOOD	GOOD	GOOD	-	V. POOR	V. POOR	POOR	GOOD	V. POOR	-

------- POTENTIAL NATIVE PLANT COMMUNITY (RANGELAND OR FOREST UNDERSTORY VEGETATION) -------

COMMON PLANT NAME	PLANT SYMBOL (NLSPN)	PERCENTAGE COMPOSITION (DRY WEIGHT) BY CLASS DETERMINING PHASE				

POTENTIAL PRODUCTION (LBS./AC. DRY WT):					
FAVORABLE YEARS					
NORMAL YEARS					
UNFAVORABLE YEARS					

--- FOOTNOTES ---

A   ESTIMATES OF ENGINEERING PROPERTIES ARE BASED ON TEST DATA FROM SIMILAR SOILS
B   RATINGS BASED ON GUIDE FOR INTERPRETING ENGINEERING USES OF SOILS, NOV. 1971
1   SEASONAL WATER TABLE WITHIN 5 FT. DEPTH BUT CONSIDERED A SLIGHT LIMITATION FOR THESE USES
C   RATINGS BASED ON NORTHEAST REGIONAL CRITERIA, MAR. 1966
D   RATINGS BASED ON SOILS MEMOS 69, OCT. 1968; 26, SEPT. 1967; OR 74, JAN. 1972

Rating / Key Phrase	Septic Tank Absorption Fields	Sewage Lagoons	Sanitary Landfills (Trench)	Sanitary Landfills (Area)	Daily Cover for Landfills	Shallow Excavations	Dwellings Without Basements	Dwellings With Basements	Small Commercial Buildings	Local Roads and Streets	Roadfill	Sand	Gravel	Topsoil	Pond Reservoir Areas	Embankments, Dikes, and Levees	Excavated Ponds, Aquifer Fed	Drainage	Irrigation	Terraces and Diversions	Grassed Waterways	Camp Areas	Picnic Areas	Playgrounds	Paths and Trails	Lawns, Landscaping
Rating																										
Slight	X	X	X	X		X	X	X	X	X					X	X	X					X	X	X	X	X
Moderate	X	X	X	X		X	X	X	X	X					X	X	X					X	X	X	X	X
Severe	X	X	X	X		X	X	X	X	X					X	X	X					X	X	X	X	X
Probable												X	X													
Improbable												X	X													
Good					X						X			X												
Fair					X						X			X												
Poor					X						X			X												
Features Affecting															X	X	X	X	X	X	X					
Key Phrases																										
Area Reclaim					X						X			X												
Cemented Pan	X	X	X			X	X	X	X	X					X			X	X	X				X		
Complex Slope																		X	X	X						
Cutbanks Cave						X												X	X							
Deep to Water																	X									
Depth to Rock	X	X	X			X	X	X	X	X					X	X	X	X		X				X		
Droughty																			X	X						
Dusty																						X	X	X	X	
Erodes Easily																X										
Excess Fines												X	X													
Excess Humus	X	X	X	X	X	X	X	X	X	X	X	X	X	X	X	X		X				X	X	X	X	X
Excess Lime														X				X								
Excess Salt														X		X		X	X	X						X
Excess Sodium														X				X	X	X						X
Fast Intake																			X							
Favorable															X	X	X	X	X	X	X					
Floods	X	X	X	X		X	X	X	X	X								X	X			X	X	X	X	X
Fragile																						X	X	X	X	
Frost Action							X	X	X	X	X															
Hard to Pack					X											X										
Large Stones	X	X	X		X	X	X	X	X	X	X	X	X	X	X	X	X	X	X	X	X	X	X	X	X	X
Low Strength							X	X	X	X	X					X										
No Water																	X									
Not Needed																		X		X	X					
Percs Slowly	X																	X	X	X	X	X		X		X
Permafrost	X	X	X	X		X	X	X	X	X					X	X				X						
Piping																X				X	X					
Pitting	X	X	X	X		X	X	X	X	X					X			X	X							
Poor outlets																		X	X							
Rooting Depth																				X	X	X				
Salty Water																			X							
Seepage		X	X	X	X										X	X		X								
Shrink-swell							X	X	X	X	X					X										
Slippage	X		X			X	X	X	X	X					X					X						
Slope	X	X	X	X	X	X	X	X	X	X	X				X	X		X	X	X	X	X	X	X	X	X
Slow Intake																			X							
Slow Refill																	X									
Small Stones		X	X		X	X								X						X		X	X	X	X	X
Soil Blowing																			X	X	X	X	X	X	X	
Subsides	X		X			X	X	X	X									X								
Thin Layer					X						X	X	X	X		X				X	X					
Too Acid					X									X						X	X	X	X	X		X
Too Clayey			X		X	X								X								X	X	X	X	
Too Sandy			X		X									X						X		X	X	X	X	
Unstable Fill		X				X	X	X	X									X								
Wetness		X	X	X	X	X	X	X	X	X	X				X			X	X	X	X	X	X	X	X	X

TABLE 20/*Soil survey interpretations for septic tank absorption fields (adapted from Soil Survey Staff, 1978).*

PROPERTY	LIMITS			RESTRICTIVE FEATURE
	SLIGHT	MODERATE	SEVERE	
1. USDA TEXTURE	—	—	ICE	PERMAFROST
2. FLOODING	NONE, PROTECTED	RARE	COMMON	FLOODS
3. DEPTH TO BEDROCK (IN)	> 72	40–72	< 40	DEPTH TO ROCK
4. DEPTH TO CEMENTED PAN (IN)	> 72	40–72	< 40	CEMENTED PAN
5. DEPTH TO HIGH WATER TABLE (FT)	—	—	+	PONDING
	> 6	4–6	0–4	WETNESS
6. PERMEABILITY (IN/HR) (24–60 IN)	2.0–6.0	0.6–2.0	< 0.6	PERCS SLOWLY
ALL LAYERS BELOW 24 IN	—	—	> 6.0	POOR FILTER
7. SLOPE (PCT)	0–8	8–15	> 15	SLOPE
8. FRACTION > 3 IN (WT PCT)	< 25	25–50	> 50	LARGE STONES

TABLE 21/*Soil survey interpretations for dwellings with basements (adapted from Soil Survey Staff, 1978).*

PROPERTY	LIMITS			RESTRICTIVE FEATURE
	SLIGHT	MODERATE	SEVERE	
1. USDA TEXTURE	—	—	ICE	PERMAFROST
2. FLOODING	NONE, PROTECTED	—	RARE, COMMON	FLOODS
3. DEPTH TO HIGH WATER TABLE (FT)	—	—	+	PONDING
	6	2.5–6	0–2.5	WETNESS
4. DEPTH TO BEDROCK (IN)				DEPTH TO ROCK
HARD	> 60	40–60	< 40	
SOFT	> 40	20–40	< 20	
5. DEPTH TO CEMENTED PAN (IN)				CEMENTED PAN
THICK	> 60	40–60	< 40	
THIN	> 40	20–40	< 20	
6. SLOPE (PCT)	0–8	8–15	> 15	SLOPE
7. SHRINK-SWELL	LOW	MODERATE	HIGH	SHRINK-SWELL
8. UNIFIED (BOTTOM LAYER)	—	—	OL, OH, PT	LOW STRENGTH
9. FRACTION > 3 IN (WT PCT)	< 25	25–50	> 50	LARGE STONES

TABLE 22/Soil survey interpretations for local roads and streets (adapted from Soil Survey Staff, 1978).

| PROPERTY | LIMITS | | | RESTRICTIVE FEATURE |
	SLIGHT	MODERATE	SEVERE	
1. USDA TEXTURE	—	—	ICE	PERMAFROST
2. DEPTH TO BEDROCK (IN)				DEPTH TO ROCK
HARD	> 40	20–40	< 20	
SOFT	> 20	< 20	—	
3. DEPTH TO CEMENTED PAN (IN)				CEMENTED PAN
THICK	> 40	20–40	< 20	
THIN	> 20	< 20	—	
4. AASHTO GROUP INDEX NUMBER	0–4	5–8	> 8	LOW STRENGTH
5. AASHTO	—	A-4, A-5	A-6, A-7, A-8	LOW STRENGTH
6. DEPTH TO HIGH WATER TABLE (FT)	—	—	+	PONDING
	> 2.5	1.0–2.5	0–1.0	WETNESS
7. SLOPE (PCT)	0–8	8–15	> 15	SLOPE
8. FLOODING	NONE, PROTECTED	RARE	COMMON	FLOODS
9. POTENTIAL FROST ACTION	LOW	MODERATE	HIGH	FROST ACTION
10. SHRINK-SWELL	LOW	MODERATE	HIGH	SHRINK-SWELL
11. FRACTION > 3 IN (WT PCT)	< 25	25–50	> 50	LARGE STONES

TABLE 23/Soil survey interpretations for topsoil (adapted from Soil Survey Staff, 1978).

| PROPERTY | LIMITS | | | RESTRICTIVE FEATURE |
	GOOD	FAIR	POOR	
1. DEPTH TO BEDROCK (IN)	> 40	20–40	< 20	AREA RECLAIM
2. DEPTH TO CEMENTED PAN (IN)	> 40	20–40	< 20	AREA RECLAIM
3. DEPTH TO BULK DENSITY > 1.8 (G/CC) (IN)	> 40	20–40	< 20	AREA RECLAIM
4. USDA TEXTURE (0–40 IN)	—	LCOS, LS, LFS, LVFS	COS, S, FS, VFS	TOO SANDY
5. USDA TEXTURE (0–40 IN)	—	SCL, CL, SICL	SIC, C, SC	TOO CLAYEY
6. USDA TEXTURE (0–40 IN)	—	—	FB, HM, SP, MPT, MUCK, PEAT, CE	EXCESS HUMUS
7. FRACTION > 3 IN (WT PCT)				
(0–40 IN)	< 5	5–25	> 25	LARGE STONES
(40–60 IN)	< 15	15–30	> 30	AREA RECLAIM
8. COARSE FRAGMENTS (PCT)				
(0–40 IN)	< 5	5–25	> 25	SMALL STONES
(40–60 IN)	< 25	25–50	> 50	AREA RECLAIM
9. SALINITY (MMHOS/CM) (0–40 IN)	< 4	4–8	> 8	EXCESS SALT
10. LAYER THICKNESS (IN)	> 40	20–40	< 20	THIN LAYER
11. DEPTH TO HIGH WATER TABLE (FT)	—	—	< 1	WETNESS
12. SODIUM ADSORPTION RATIO (GREAT GROUP)	—	—	> 12 (HALIC, NATRIC, ALKALI PHASES)	EXCESS SODIUM
13. SOIL REACTION (pH) (0–40 IN)	—	—	< 3.6	TOO ACID
14. SLOPE (PCT)	0–8	8–15	> 15	SLOPE

TABLE 24/_Soil survey interpretations for pond reservoir area (adapted from Soil Survey Staff, 1978)._

PROPERTY	LIMITS			RESTRICTIVE FEATURE
	SLIGHT	MODERATE	SEVERE	
1. USDA TEXTURE	—	—	ICE	PERMAFROST
2. PERMEABILITY (IN/HR) (20–60 IN)	< 0.6	0.6–2.0	> 2.0	SEEPAGE
3. DEPTH (IN) TO LAYER WITH PERM ⩾ 2.0	> 60	40–60	< 40	SEEPAGE
4. DEPTH TO BEDROCK (IN)	> 60	20–60	< 20	DEPTH TO ROCK
5. DEPTH TO CEMENTED PAN (IN)	> 60	20–60	< 20	CEMENTED PAN
6. SLOPE (PCT)	< 3	3–8	> 8	SLOPE
7. USDA TEXTURE	—	—	MARL, GYP	SEEPAGE

TABLE 25/_Soil survey interpretations for drainage (adapted from Soil Survey Staff, 1978)._

PROPERTY	LIMITS	RESTRICTIVE FEATURE
1. USDA TEXTURE	ICE	PERMAFROST
2. DEPTH TO HIGH WATER TABLE (FT)	> 3 +	DEEP TO WATER PONDING
3. PERMEABILITY (IN/HR) (0–40 IN)	< 0.2	PERCS SLOWLY
4. DEPTH TO BEDROCK (IN)	< 40	DEPTH TO ROCK
5. DEPTH TO CEMENTED PAN (IN)	< 40	CEMENTED PAN
6. FLOODING	COMMON	FLOODS
7. TOTAL SUBSIDENCE	ANY ENTRY	SUBSIDES
8. FRACTION > 3 IN (WT PCT)	> 25	LARGE STONES
9. POTENTIAL FROST ACTION	HIGH	FROST ACTION
10. SLOPE (PCT)	> 3	SLOPE
11. USDA TEXTURE	COS, S, FS, VFS, LCOS, LS, LFS, LVFS, SG, G	CUTBANKS CAVE
12. SALINITY (MMHOS/CM)	> 8	EXCESS SALT
13. SODIUM ADSORPTION RATIO (GREAT GROUP)	> 12 (NATRIC, HALIC)	EXCESS SODIUM
14. SULFIDIC MATERIALS (GREAT GROUP)	SULFAQUENTS, SULFIHEMISTS	EXCESS SULFUR
15. SOIL REACTION (PH)	< 3.6	TOO ACID
16.	NONE OF ABOVE	FAVORABLE

TABLE 26/_Soil survey interpretations for irrigation (adapted from Soil Survey Staff, 1978)._

PROPERTY	LIMITS	RESTRICTIVE FEATURE
1. FRACTION > 3 IN (WT PCT)	> 25	LARGE STONES
2. DEPTH TO HIGH WATER TABLE (FT)	< 3 +	WETNESS PONDING
3. AVAILABLE WATER CAPACITY (IN/IN)	< 0.10	DROUGHTY
4. USDA TEXTURE (SURFACE LAYER)	S, FS, VFS, LS LFS, LVFS	FAST INTAKE
5. USDA TEXTURE (SURFACE LAYER)	SIC, C, SC	SLOW INTAKE
6. WIND ERODIBILITY GROUP	1, 2, 3	SOIL BLOWING
7. PERMEABILITY (IN/HR) (0–60 IN)	< 0.2	PERCS SLOWLY
8. DEPTH TO BEDROCK (IN)	< 40	DEPTH TO ROCK
9. DEPTH TO CEMENTED PAN (IN)	< 40	CEMENTED PAN
10. FRAGIPAN (GREAT GROUP)	ALL FRAGI	ROOTING DEPTH
11. BULK DENSITY (G/CC) (0–40 IN)	> 1.7	ROOTING DEPTH
12. SLOPE (PCT)	> 3	SLOPE
13. EROSION FACTOR (K) (SURFACE LAYER)	> 0.35	ERODES EASILY
14. FLOODING	COMMON	FLOODS
15. SODIUM ADSORPTION RATIO (GREAT GROUP)	> 12 (NATRIC, HALIC)	EXCESS SODIUM
16. SALINITY (MMHOS/CM)	> 8	EXCESS SALT
17. SOIL REACTION (PH)	< 3.6	TOO ACID
18.	NONE OF ABOVE	FAVORABLE

TABLE 27/*Soil survey interpretations for playgrounds (adapted from Soil Survey Staff, 1978).*

PROPERTY	LIMITS SLIGHT	LIMITS MODERATE	LIMITS SEVERE	RESTRICTIVE FEATURE
1. USDA TEXTURE	—	ST	STV, STX, BYV, BYX, CB, CBV, FL, FLV, BY	LARGE STONES
2. SLOPE (PCT)	0–2	2–6	> 6	SLOPE
3. COARSE FRAGMENTS (PCT) (SURFACE LAYER)	< 10	10–25	> 25	SMALL STONES
4. USDA TEXTURE (SURFACE LAYER)	—	—	SC, SIC, C	TOO CLAYEY
5. USDA TEXTURE (SURFACE LAYER)	—	LCOS, VFS	COS, S, FS	TOO SANDY
6. UNIFIED (SURFACE LAYER)	—	—	OL, OH, PT	EXCESS HUMUS
7. DEPTH TO HIGH WATER TABLE (FT)	> 2.5 —	1.5–2.5 —	< 1.5 +	WETNESS PONDING
8. FLOODING	NONE, RARE, PROTECTED	OCCAS	FREQ	FLOODS
9. DEPTH TO BEDROCK (IN)	> 40	20–40	< 20	DEPTH TO ROCK
10. DEPTH TO CEMENTED PAN (IN)	> 40	20–40	< 20	CEMENTED PAN
11. PERMEABILITY (IN/HR) (0–40 IN)	> 0.6	0.06–0.6	< 0.06	PERCS SLOWLY
12. USDA TEXTURE (SURFACE LAYER)	—	SIL, SI, VFSL, L	—	DUSTY
13. SODIUM ADSORPTION RATIO (GREAT GROUP)	—	—	> 12 (NATRIC, HALIC)	EXCESS SODIUM
14. SALINITY (MMHOS/CM)	< 4	4–8	> 8	EXCESS SALT
15. SOIL REACTION (PH) (SURFACE LAYER)	—	—	< 3.6	TOO ACID

TABLE 28/*Ratings of soils for rainfed and irrigated paddy rice.*

Item Affecting Use	Soil Potential Very good	Good	Moderate	Poor	Very poor
Drainage class	Somewhat poorly drained	Moderately well drained	Well drained	Poorly drained	Very poorly drained
Depth (cm)	> 200	100–200	50–100	20–50	< 20
Slope (%)	< 2	2–4	4–6	6–8	> 8
Texture	c sic	cl sicl	si, sil, scl	l sl	s ls
Permeability (cm/hr)	< 0.5	0.5–1.5	1.5–5.0	5.0–15.0	> 15.0
Stones and rock outcrops	None	Few	Some	Many	Very many
Puddling qualities	OL, OH	ML, CL, MH, CH	GM, GC, SM, SC	GW, GP SW, SP	Pt
Salinity (mmhos/cm)	< 2	2–4	4–8	8–10	> 10
Erosion	None	Slightly eroded	Moderately eroded	Severely eroded	Very severely eroded
Flooding	None	Rare	Occasional	Frequent	Very frequent

Projects

The best format for teaching soil survey interpretations is through projects, where students can learn by doing. Teachers of courses involving soil survey interpretations should first concentrate on teaching the basic language of soil survey interpretations (description, definition, mapping, analysis) and the criteria by which soils are rated for different purposes (Soil Taxonomy, computerized groupings of soils). Then students should select an individual project to pursue through the duration of the course. It is good practice to ask each student to select a particular project as early as possible in the course, so that each person has ample time to think about the topic, to research various aspects of the problem, and to contact various people involved with making and using soil surveys.

After presenting the material in Part I of this field guide to the class (before the first exam), students should be asked to select an individual project topic. Guest lectures should be delivered to the class by at least two people involved with using soil surveys in an official capacity (e.g., SCS District Conservationist, County Planner) to give different perspectives on uses of soil surveys. Students should be encouraged to pick topics relevant to the needs of the community. An effective device to make the project relevant for each student is for the teacher to arrange to deposit the project reports (at the end of the course after grading) in the library of the County Planning Office—where they can be used by planners, engineers, and the general public. A typed report of at least 30 pages should be required, and the students should be informed as early in the course as possible about the exact style for writing and typing and filing the report. The project report should constitute at least half the grade for the course.

When this course is taught in full format at Cornell University, students are given a list of possible projects (Table 29) as suggestions and asked to select a topic of their own choosing. They are also given recent newspaper clippings about soils-related problems in the community and elsewhere (Table 30). The lists give suggestions only, and each student has complete freedom in selecting a project topic as long as the topic deals with the subject matter of the course (soil survey interpretations). Some undergraduate students without previous courses in soils may select a very simple project of merely coloring a few maps and writing a brief report; other students at a graduate level may select a topic that contributes to their Ph.D. thesis. The teacher works closely with each student individually throughout the course on each project topic. Experience has shown that most students become so interested in their project (of their own choosing) that they devote far more time and energy to it than is required. Table 31 lists projects done by a recent class at Cornell University together with grades and some of the evaluation comments. Typically, a few reports are outstanding, most are good, and a few are not so good. The teacher's detailed comments of evaluation are also deposited with the planning department, to help the users evaluate the merits of each report. Many of these reports have proved to be extremely useful to the community and some have stimulated additional student and professional staff projects. Over the years, a library of student reports has been accumulated, so that current classes can consult the work of past classes and improve or expand on the past projects.

Under the project format, each student has the role of an outside consultant researching a soil-related problem according to the needs of the community. Different local experts are consulted. Detailed soil maps and reports are interpreted to make the information most understandable to the user. Highly contrasting soils of a limited area are examined and interpreted for a specific use or for multiple uses. Photographs are taken in the area of investigation and some fieldwork is accomplished. At the conclusion of the course, the student submits a typed report and gives an oral presentation of the work. Through the oral talk to the class, each student presents his or her own ideas, and each listening student learns from all of the others. Initiative and imagination are encouraged. In the evaluations of the course, students consistently list the project as one of the best parts of the course. Learning by doing in the project format is extremely effective in teaching and learning about soil survey interpretations.

TABLE 29/*List of possible class projects for individuals in soil survey interpretations and maps.*

Runoff from different soils
Pollution from different soils
Diseases and mortality rates on different soils
Alternative soils for road route selections
Unique soils for preservation
Aesthetics of soils
Capacity of soils for animal waste disposal
Suitability of soils for subdivisions
Location and characteristics of sand and gravel sources under soils
Geologic history of different soils and implications to uses of areas
Geomorphological groups of soils influencing use
Climates of soils and use implications
Relationships between soils and land use via computers
History of uses of soils—historical aspects of Tompkins County or New York State
Human activities in and on different soils
Capability groups of soils
Agricultural yields of soils
Woodland suitability groups of soils
Wildlife suitability groups of soils
Engineering classifications of soils
Limitations of soils for homesites
Limitations of soils for streets and parking lots in subdivisions
Limitations of soils for sanitary landfill
Limitations of soils for pipeline installations

Strategic implications of soil properties for military purposes—past, present, or future
Limitations of soils for lawns, landscaping, and golf fairways
Limitations of soils for campsites
Limitations of soils for play and picnic areas (extensive use)
Limitations of soils for athletic fields
Likely sources of topsoil in different soils
Location of floodplains (soils formed in alluvial sediments)
Well yields in geologic materials beneath different soils
Alternative uses of different soil areas
Competition for use of different soil areas
Cost of structures in different soils
Performances of structures in different soils
Attitudes and opinions of people living on different soil resources
Improving designs of structures in different soils
Aesthetic potential of different soils
Governmental regulations of soil use
Improper judgment in uses of soils
Highway right-of-way design using soils information
Benefits from using soil surveys
How to improve soil profile descriptions
Tree growth in soil map units
TDN (total digestible nutrients) yields from different soils
Rational taxes based on soils

TABLE 30/*List of newspaper articles given to class on "Use of Soil Information and Maps as Resource Inventories" to help students select project topics relating to solving soil-related problems of the community.*

"Proposed connector from Rt.13 to Rt. 79 meets opposition." *The Ithaca Journal*, 28 July 1981.
"Growth and decline (of counties) in the state's population 1970 vs. 1980." *The New York Times*, 26 July 1981.
"Ideal Farms sells off its dairy herd." *The New York Times*, 16 May 1981.
"New York Area's reliance on imported food grows." *The New York Times*, 31 July 1981.
"Produce farmers resurging in New York State." *The New York Times*, 31 August 1981.
"EMC (Environmental Management Council) launches search for county's wetlands." *The Ithaca Journal*, 25 May 1981.
"Local counties baffled by landfill rules." *The Ithaca Journal*, 3 June 1981.
"Some Tompkins County farmers will get smaller tax break than expected." *The Ithaca Journal*, 23 April 1981.
"Sewer drainage troubles Moravia." *The Ithaca Journal*, 22 July 1981.
"Farm governors seek federal help on soil erosion." *The Ithaca Journal*, 11 August 1981.
"A state of emergency in vulnerability of Pennsylvania's food system." *The New York Times*, 7 June 1981.
"America must preserve its farmland." *The Ithaca Journal*, 18 May 1981.
"Flooding leaves hundreds homeless (in Montana)." *The Ithaca Journal*, 25 May 1981.
"Famine confronts 150 million in Africa." *The New York Times*, 26 March 1981.
"Old mingles with new in a sandy Nigerian city." *The New York Times*, 28 July 1981.
"Why give away the granary?" (Editorial) *The New York Times*, 24 April 1981.

TABLE 31/*List of student project reports and evaluations on file in the library of the planning department from one-semester teaching of the course "Use of Soil Information and Maps as Resource Inventories."*

"The Limitations of the Soils Surrounding Dryden for Land Application of Wastes"—A—Good map—Howard and Phelps soils in NE north of Route 38 offer some promise.

"Soil Erosion Research Area in Cornell University Animal Teaching and Research Center"—B—Difficult to understand—Photos would have helped.

"An Attempt to Characterize Forest-Soil Relationships Based on Map Information"—C—Spelling terrible—Three typed pages and three overlay maps unbound—Too brief.

"The Aesthetic Potential of Soils"—A—Beautiful colored map of Lansing—Writing good—Typing neat—Good illustration of imaginative applications of soil survey.

"The Influence of Soil Conditions on Agricultural Viability in Caroline"—A—Effective photos—Useful report.

"Recreational Use of Soils on Cornell University Properties"—A—Photos shock and delight hikers along Finger Lakes trail.

"Mined-land Use Planning"—A—Good contribution toward solving problems—Lansing gravel pit reclamation.

"The Development Potential of the Savage Farm"—A—Beautiful folder layout—Soil problems well defined—Plenty of illustrations.

"Water Tables and Septic Tanks"—B+—Report not well integrated—Text not coordinated with appendix and figures.

"Feasibility of Extension to Tompkins County Airport Runway"—A—Attractive and neat report.

"Soil Suitability for Sanitary Landfills"—B+—Lifted tables inadequately integrated into report.

"Characterization of Erosion in the Hills of Ithaca"—A—Good report submitted early!

"Soil Limitations and Construction in the Town of Lansing"—A++—Marvelous report!

"Prime Farmland Soils of Tompkins County"—A—Report elegant in its simplicity.

"Conflicting Uses of Soils in Sapsucker Woods"—B+—Illustrated how lack of planning is degrading the environment for birds as well as people.

"Play and Picnic Areas in Caroline"—A—Report needs photos—Neatly typed.

"Agricultural Soil Use in Ulysses"—B—Typing smudged.

"Frost Susceptible Soils in Lansing"—A—Good mastery of the engineering aspects of frost heaving.

"A Rating System of Soils for Development in Ithaca"—A—Good innovative and imaginative report.

"Fall Creek Suspended Sediment"—A—Good report contains data from monitored stations.

"Road Bank Erosion in Tompkins County"—B—First-person narration in report is not good.

"Soil Information in Selection of Woody Plants in Ithaca"—A—Good report that could save thousands of dollars.

"Gravel Extraction in Ulysses and Enfield"—B+—Report submitted late.

Photographs

PURPOSE Photographs are probably the best tool to illustrate effects of soil characteristics on land use, especially for laypersons. The purpose of this exercise is to assign students to take photos to illustrate soil conditions in their study areas or project areas, and to show examples of some photographs that illustrate principles of soils perspectives. Each student should take at least 50 35-mm black-and-white or more than 20 color photos of contrasting soil properties within his or her area of interest. These photos should be taken over a period of at least several days or weeks so that differences in weather and lighting and angle of sun can be illustrated. Take several photos of the same scene under different weather conditions (calm versus windy; cloudy versus sunny; midday versus morning and evening; etc.). Experiment with close-up photos and wide-angle and telephoto lenses at different shutter speeds. Make notes on camera settings and weather conditions when each photo is taken. A lecture period should be devoted to discussion of the results, with examples exhibited of work of the various students. This exercise is an excellent method of learning by doing.

EXAMPLES The following photographs illustrate some of the principles involved. Figure 30 is a close-up of roots of eucalypts that have been exposed by wind erosion. By getting close to the subject and carefully selecting the right angle, the photographer was able to emphasize the aesthetics of the weird and unusual formations made by the exposed roots. The shady dark side of the roots emphasizes the shape in shadows against the bright sky. The strange shapes force the viewer to think about the subject, which is an extensive area where more than 1 meter of soil has been removed by land abuse which caused the soil loss. Overgrazing is probably the cause, and the viewer must surely ponder about what can be done to reclaim the devastated landscape. Not even an impoverished sheep or goat is visible in the distance. If the photographer had taken a more distant view, the impact of the erosion process would probably have been much less.

Aerial photographs at an oblique angle often can be used to show landscape processes and landform shapes. Figure 31 illustrates excellently the erosion and deposition on bare erosive soils. Notice that the aircraft from which the photo was taken is at a low altitude, and that the sun angle is low (morning or evening). The long shadows accentuate the shapes of the shallow rills and small gullys. The plume of smoke in the distance adds to the viewer's feeling of devastation of the landscape. The photograph gives a feeling of desolation and desperation to the viewer, and anyone who has lived on such a farm knows the mental and physical anguish of such an experience of crops and soils washed away. This photograph presents a powerful argument for soil erosion control on farms.

Perspective is important, and the photographer must become a part of the landscape and compose the scene from his or her experience to create the desired effect. Figure 32 shows a village road in Honduras, with terrible ruts in the road (street) in the foreground. As a matter of fact, the photographer's vehicle was stuck in the mud when the photo was taken. This photo conveys all of the implications of the horrifying effects of poor transportation systems due to poor soil conditions and lack of proper road construction and maintenance. Sick and injured people in the village cannot be transported quickly to a hospital, crops and produce cannot be effectively moved to market, village shops cannot be efficiently stocked, people cannot conveniently take a bus to the city to visit relatives, and the whole feeling of isolation and poverty is conveyed to the viewer. This photo helps the reader to realize that soil surveys and their interpretations should be of top priority toward regional economic development in a country. The human condition is closely correlated to the soil conditions; this scene is near refugee camps and areas of revolutionary activities.

Photographers of soil conditions must be ever-perceptive and intensely seeking for scenes that express even the subtle expression of characteristics of soils. Figure 33 illustrates a modern scene at a shopping center in Manila. Most observers would see nothing wrong with the soils in the photo. On examining the photo closely, however, a definite warping can be seen in the surface of the parking lot pavement. This warping is due to the Vertisol characteristics of the soils, which are prone to shrinking and swelling. The unique lighting characteristics, the thin film of water on the pavement surface, and the accumulation of moisture in shallow puddles in depressions all contribute to accentuate the microrelief. The unique expression of surface warping in this photograph was not noticed until several weeks after the photo was taken, not until after the film had been developed and printed.

Figure 34 is another scene from the Philippines where very poorly drained soils have been developed for human habitation without adequate drainage. Many disease and health problems are caused or aggravated by the wet soil conditions. Also, the political and economic conditions here are closely correlated to the soils. Often, swamps are developed because the land is cheap and can be sold after construction for quick profit. Or exploitive landlords capitalize on the population pressures with high rental charges. Commonly, areas such as this are the scenes of guerilla antigovernment activities. Increasingly, progressive governments are recognizing that the human condition in such places must be improved to achieve a stable society, and programs are instituted to improve conditions by providing drainage and improving the soils.

Some problems of soil management are truly overwhelming, and must be publicized by photographs and other methods so that government action can be encouraged to take place as soon as possible. In Bangkok, for example, rapid spread of urbanization and industrialization is causing excessive pumping of the groundwater from unstable geologic substrata. With the removal of the groundwater, the soil surface subsides—as much as 4 inches per year in some areas. The problem is increased because Bangkok is located on a low-lying delta at an elevation close to sea level, the area has a pronounced rainy monsoon season, and the sea level is rising because of the melting and recession of the polar ice caps. Scenes such as that in Figure 35 illustrate the problem, but flooding and subsidence is not apparent to most people during much of the year. Thus, most people are unaware

of the problem and the need for immediate action, and only a few scientists know the exact measurements of the subsidence (Figure 36). Control of the problem would involve legislation to regulate groundwater pumping, increasing supply of water from other sources, construction of dikes and pumping stations, building dams and flood control structures at the upper reaches of the watershed, and reducing runoff and soil erosion throughout the vast drainage basin. Soil surveys should provide an important base in planning and development of all these land-use improvements. Photographs such as those in Figures 35 and 36 can play an important role in informing the public and stimulating government action. Such soil subsidence problems are common to many low-lying coastal cities, including London, New Orleans, and Venice.

Contrast photographs can be especially valuable to show good and poor soil conditions, and the resultant economic consequences. In New York State 100 years ago, most of the land was occupied by small farms, regardless of the soil conditions. Steep slopes and wet soils were plowed with teams of horses, and even stony and shallow-to-bedrock places were cleared and cultivated. Soil surveys were not available, and land settlement was by trial and error. After 100 years, however, the results of the trial-and-error experiment of land development can be seen scattered over the landscape. Soils with steep slopes, poorly drained conditions, shallowness to bedrock, infertility, and other severe problems are generally related to abandoned or failing farms—and the good soils are companions with prosperous farms and productive agriculture. Statistically, the better farms are located on the better soils, and the poorer farms are located on the poorer soils (with only a few exceptions). Figures 37 and 38 are paired contrast photographs that illustrate the corresponding soil and land-use correlations. Both photos were taken within the same area near Lowville in northern New York State. Such contrast photographs present a powerful argument that soils are important in determining land use, and that we should be more aware of soil differences in the future than we have been in the past.

FIGURE 30/*Roots of mallee (eucalypt) thicket or brush vegetation which have been exposed by wind erosion of the surrounding soil in northwestern Victoria, Australia (photo from Soil Conservation Authority).*

FIGURE 31/*Oblique aerial photograph of eroded farm near Walla Walla, Washington, after a heavy rain (photo by Soil Conservation Service).*

FIGURE 32/*Ruts in road in Honduras in soils with a critical plasticity index (photo by David Olson).*

FIGURE 33/*Warping of pavement in Manila according to the expansion and contraction of the soils.*

FIGURE 34/*Urban development on very poorly drained soils. This area could be much improved for human habitation through drainage.*

FIGURE 35/*Flooding in Bangkok due to subsidence of the soils. Excessive groundwater pumping is lowering the geologic substrata, and the sinking soil surface increases the flood hazards with each passing year.*

FIGURE 37/*Abandoned farm on soils with problems for agriculture, near Lowville in northern New York State.*

FIGURE 36/*Monitor well installed to a depth of several hundred feet a few months before this photo was taken. The elevation of the tube casing is stable, but the concrete floor in the shelter for the well has broken away as the ground surface settles. This photo provides a definite record that the soil surface has subsided several inches within just a few months time.*

FIGURE 38/*Productive farm on soils with good characteristics for farming, near Lowville in northern New York State.*

First exam

The first exam is designed as a "learning experience" covering the first half of the textbook and the field guide content to this point. This format for a first exam is one that can be adapted to many teaching situations. To prepare for the exam, students are informed about the general nature of each question—but not the exact specifics. Question 1 asks the student to describe the project that he or she plans to do in soil survey interpretations. Question 2 measures the student comprehension of the completeness of the subject matter of the soil survey. Question 3 tests knowledge of the definitions. Question 4 and 5 evaluate the student's ability to transfer soil profile descriptions for two highly contrasting soils into ratings for different uses. All items in this "model" first exam can be revised and modified to fit the class and the local environment. Thus, local soil profile descriptions (or descriptions from the student study area) should be used. Terms to be defined can be changed to fit each teacher need. When finished with the exam, each student should feel that he or she has a good grasp of the subject matter of soil survey interpretations—and should be ready to start on the project and the other aspects of interpretations about uses of soil surveys.

FIRST EXAM

(Be as comprehensive as you can in answering these questions.)

1. Outline below what you would like to do as a project of soil survey interpretations in your area. Tell how you will use the information available in the soil survey and gather additional information to assist in planning specific uses of soils. Tell how you will coordinate and supplement the work by coordination with the District Conservationist of the Soil Conservation Service, the local planner, the Cooperative Extension Agent, or others who might be interested in the project. You should have already studied some of the project reports that previous students have completed.

PROJECT DESCRIPTION _____

_____

_____

_____

_____

_____

_____

_____

_____

_____

_____

_____

_____

_____

_____

2. List soil characteristics and properties described in a soil survey.

_____

_____

_____

_____

_____

_____

_____

_____

_____

_____

_____

_____

_____

_____

_____

3. Briefly define the following terms.

Hue _____

Value _____

Chroma _____

Sand _____

Silt _____

Clay _____

Grade of structure _____

Type of structure _____

Wet consistence _____

Abrupt boundary _____

Gradual boundary _____

Horizon_____

Mottle _____

Munsell color designation _____

Reaction _____

Abundance of mottles_____

Texture _____

Subordinate horizon symbols _____

Cobbles _____

Gravel _____

Common medium distinct mottles _____

Coarse moderate platy structure _____

II B23t horizon _____

4. The soil described briefly below occupies a significant acreage in your area. What are the general suitabilities or limitations of the soil for corn, red pine, excavated impoundments (dug ponds), trafficability of heavy wheeled vehicles, septic tank seepage fields, building foundations, sanitary landfill (garbage burial), campsites, and why? What are the likely yields of wells drilled into groundwater aquifers beneath the soil, and why? (Assume nearly level "A" slopes for this soil.)

Ap—0 to 9 inches; dark brown (10YR 3/3) fine sandy loam; weak fine granular structure; friable; many roots; 10 percent coarse fragments; strongly acid; clear smooth boundary.
B21—9 to 18 inches; yellowish brown (10YR 5/6) gravelly loam; weak medium subangular blocky structure; friable; common roots; many pores; 20 percent coarse fragments; strongly acid; clear smooth boundary.
B22—18 to 25 inches; yellowish brown (10YR 5/6) gravelly sandy loam; weak medium gran-
ular structure; friable; few roots; few pores; 20 percent coarse fragments; very strongly acid; abrupt wavy boundary.
C1—25 to 42 inches; light olive brown (2.5Y 5/4) gravelly sandy loam; massive; very friable; few pores; 30 percent coarse fragments; strongly acid; clear wavy boundary.
C2—42 to 60 inches; light olive brown (2.5Y 5/4) very gravelly sandy loam; massive; very friable; 55 percent coarse fragments; medium acid.

Use	Suitability or Limitation	Reason(s)
Corn		
Red pine		
Excavated impoundments (dug ponds)		
Trafficability		
Septic tanks		
Foundations		
Sanitary landfills		
Campsites		
Well yields		

5. The soil described briefly below occupies a significant acreage in your area. What are the general suitabilities or limitations of the soil for corn, red pine, excavated impoundments (dug ponds), trafficability of heavy wheeled vehicles, septic tank seepage fields, building foundations, sanitary landfill (garbage burial), campsites, and why? What are the likely yields of wells drilled into groundwater aquifers beneath the soil, and why? (Assume nearly level "A" slopes for this soil.)

Ap—0 to 9 inches; very dark grayish brown (10YR 3/2) silt loam, grayish brown (10YR 5/2) dry; moderate fine granular structure; friable; many roots; medium acid; clear wavy boundary.

B2—9 to 17 inches; dark gray (10YR 4/1) silt loam; many medium distinct grayish brown (10YR 5/2) mottles; moderate fine granular structure; friable; common roots; medium acid; clear smooth boundary.

C1g—17 to 35 inches; olive gray (5Y 5/2) silt loam; common distinct gray (5Y 6/1) and common medium prominent yellowish brown (10YR 5/6) mottles; weak medium subangu-lar blocky structure; friable; few roots; common pores; medium acid; gradual smooth boundary.

C2g—35 to 47 inches; light olive gray (5Y 6/2) silt loam; common medium prominent yellowish brown (10YR 5/6) and common medium faint gray (5Y 6/1) mottles; weak medium subangular blocky structure; friable; few roots; common pores; medium acid; clear smooth boundary.

C3g—47 to 60 inches; gray (10 YR 5/1) fine sandy loam; common coarse distinct strong brown (7.5YR 5/6) mottles; massive; friable; few roots; medium acid.

Use	Suitability or Limitation	Reason(s)
Corn		
Red pine		
Excavated impoundments (dug ponds)		
Trafficability		
Septic tanks		
Foundations		
Sanitary landfills		
Campsites		
Well yields		

II | Applications of soil surveys in systems of wide usage

Engineering applications

PURPOSE Engineering uses of soils affect our lives every day. This exercise is designed to encourage study of different behaviors of soils subject to various construction activities. Site and regional development is very much dependent on the building of roads, houses, factories, stores, airports, embankments, and the uses of soils as source material for gravel, sand, and roadfill. In preparation for this exercise, study pages 46–48 and 91–95 of the textbook. The references will also be valuable in determining the most appropriate engineering applications of soil surveys (Olson and Warner, 1974).

APPLICATIONS Detailed soil surveys made at a scale of about 4 inches to 1 mile show soil areas down to 1 hectare in size. Often, construction sites are smaller, so that additional extremely detailed and deeper borings are needed for design of specific foundations. Soil maps, however, are invaluable in initial planning of developments and in preliminary selection of sites where more detailed studies can occur. Route selections for roads and pipelines provide good examples where soil maps are useful. When alternative routes are considered, based on soil maps, more detailed special engineering soils studies can be made of the right-of-way. Over time, the deterioration of the roads (Figure 39) and the pipelines is directly dependent on the characteristics of the soils. The design cost to overcome the soil limitations is also dependent on the soil conditions.

With big buildings (Figure 40) standard soil maps are less useful, but initial city planning, excavations, and design of new suburbs commonly make considerable use of soil surveys (Olson and Marshall, 1968). Suburbs around New Orleans, for example, often have subsidence and wetness problems. Soil maps have proved to be of great value for planning in the area, and in design of pilings, foundations, and other engineering features to overcome the severe soil limitations (Soil Survey Staff, 1971).

Study FAO Soils Bulletin 19 (FAO, 1977) to get an idea of how engineering data mesh with soil survey data. From the National Soils Handbook (Soil Survey Staff, 1978), select engineering uses of soils that will offer high-performance contrasts for soils in your area of study. Tables 32 to 37 illustrate some engineering applications of soil surveys for shallow excavations, embankments, dikes, levees, small commercial buildings, gravel, sand, and roadfill. From Table 22 (roads and streets) and the soil map of your area, select the best routes for a road and street network into the area. Then use Table 32 to locate the best places for shallow excavations for buildings. Tables 33 and 37 will be useful to locate and design embankments and roadfill. Table 34 gives ratings of soils for small commercial buildings. Table 35 will help locate gravel for roadbed and foundation subbase, and Table 36 identifies probable and improbable sources of sand for mixing concrete. The soil names and the "Form 5" evaluations (see Tables 16 to 18) also provide ratings of the soils for various uses.

Coordinate the "engineering applications" with the other aspects of the project work in your study area. Maximum benefits of soil surveys are obtained when environmental improvements are interdisciplinary. Thus, soil surveys provide on-site information of a specified determinable reliability, in accord with the detail of the soil examinations and the scale of the map. The soil survey also provides information for planning erosion control, runoff structures, impoundments, drainage, irrigation, and other off-site information necessary to contribute to the best design of a community, an industry, a farm, or a development area.

REFERENCES

FAO. 1977. Soil survey interpretation for engineering purposes. Soils Bulletin 19, Food and Agriculture Organization of the United Nations, Rome, Italy. 24 pages.

Olson, G. W. and R. L. Marshall. 1968. Using high-intensity soil surveys for big development projects: A Cornell experience. Soil Science 105:223–231.

Olson, G. W. and J. W. Warner. 1974. Engineering soil survey interpretations. Information Bulletin 77, New York State College of Agriculture and Life Sciences, Cornell University, Ithaca, NY. 8 pages.

Soil Survey Staff. 1971. Guide for interpreting engineering

uses of soils. Soil Conservation Service, U.S. Dept. of Agriculture, U.S. Government Printing Office, Washington, DC. 87 pages.

Soil Survey Staff. 1978. National soils handbook. Part II, Section 400, Application of soil survey information,

Procedure Guide. Soil Conservation Service, U.S. Dept of Agriculture, Washington, DC. Reproduced as Cornell Agronomy Mimeo 82-15, Cornell University, Department of Agronomy, Ithaca, NY. About 100 pages.

FIGURE 39/*Road deterioration on soils high in clays and silts with a critical plasticity index. The photograph was taken in early spring in Ithaca, New York, after frost action had broken up much of the road.*

FIGURE 40/*Beam-support shoring to prevent excavation caving in foundation construction for a large building in New Orleans. Many soils in this area are extremely wet and achieve a liquid state during a significant part of the year. Many deep borings closely spaced were necessary for design of these buildings, and soil surveys have proved to be very useful for planning suburbs and building construction in the outer newer areas of expanding New Orleans (photo by Ken Olson).*

TABLE 32/*Soil survey interpretations for shallow excavations (adapted from Soil Survey Staff, 1978).*

PROPERTY	LIMITS SLIGHT	LIMITS MODERATE	LIMITS SEVERE	RESTRICTIVE FEATURE
1. USDA TEXTURE	—	—	ICE	PERMAFROST
2. DEPTH TO BEDROCK (IN)				DEPTH TO ROCK
HARD	> 60	40–60	< 40	
SOFT	> 40	20–40	< 20	
3. DEPTH TO CEMENTED PAN (IN)				CEMENTED PAN
THICK	> 60	40–60	< 40	
THIN	> 40	20–40	< 20	
4. USDA TEXTURE (20–60 IN)	—	SI	COS, S, FS, VFS, LCOS, LS, LFS, LVFS, G, SG	CUTBANKS CAVE
5. USDA TEXTURE (20–60 IN)	—	C, SIC	—	TOO CLAYEY
6. SOIL ORDER	—	—	VERTISOLS	CUTBANKS CAVE
7. BULK DENSITY (G/CC)	—	> 1.8	—	DENSE LAYER
8 UNIFIED (20–60 IN)	—	—	OL, OH, PT	EXCESS HUMUS
9. FRACTION > 3 IN (WT PCT)	< 25	25–50	> 50	LARGE STONES
10. DEPTH TO HIGH WATER TABLE (FT)	—	—	+	PONDING
	> 6	2.5–6	0–2.5	WETNESS
11. FLOODING	NONE, RARE, PROTECTED	COMMON	—	FLOODS
12. SLOPE (PCT)	0–8	8–15	> 15	SLOPE

TABLE 33/*Soil survey interpretations for embankments, dikes, and levees (adapted from Soil Survey Staff, 1978).*

PROPERTY	LIMITS SLIGHT	LIMITS MODERATE	LIMITS SEVERE	RESTRICTIVE FEATURE
1. USDA TEXTURE	—	—	ICE	PERMAFROST
2. LAYER THICKNESS (IN)	> 60	30–60	< 30	THIN LAYER
3. UNIFIED	—	—	GW, GP, SW, SP, GW–GM, GP–GM, SW–SM, SP–SM, SM, GM	SEEPAGE
4. UNIFIED	—	GM, CL	ML, SM, SP, CL–ML	PIPING
5. UNIFIED	—	—	PT, OL, OH	EXCESS HUMUS
6. UNIFIED	—	—	MH, CH	HARD TO PACK
7. FRACTION > 3 IN (WT PCT)	< 15	15–35	> 35	LARGE STONES
8. DEPTH TO HIGH WATER TABLE (FT)	—	—	+	PONDING
APPARENT	> 4.0	2.0–4.0	< 2.0	WETNESS
PERCHED	> 3.0	1.0–3.0	< 1.0	WETNESS
9. SODIUM ADSORPTION RATIO (GREAT GROUP)	—	—	> 12 (NATRIC, HALIC)	EXCESS SODIUM
10. SALINITY (MMHOS/CM)	< 8	8–16	> 16	EXCESS SALT

TABLE 34/*Soil survey interpretations for small commercial buildings (adapted from Soil Survey Staff, 1978).*

| PROPERTY | LIMITS | | | RESTRICTIVE FEATURE |
	SLIGHT	MODERATE	SEVERE	
1. USDA TEXTURE	—	—	ICE	PERMAFROST
2. FLOODING	NONE, PROTECTED	—	RARE, COMMON	FLOODS
3. DEPTH TO HIGH WATER TABLE (FT)	—	—	+	PONDING
	> 2.5	1.5–2.5	0–1.5	WETNESS
4. SHRINK-SWELL	LOW	MODERATE	HIGH	SHRINK-SWELL
5. SLOPE (PCT)	0–4	4–8	> 8	SLOPE
6. UNIFIED	—	—	OL, OH, PT	LOW STRENGTH
7. DEPTH TO BEDROCK (IN)				DEPTH TO ROCK
HARD	> 40	20–40	< 20	
SOFT	> 20	< 20	—	
8. DEPTH TO CEMENTED PAN (IN)				CEMENTED PAN
THICK	> 40	20–40	< 20	
THIN	> 20	< 20	—	
9. FRACTION > 3 IN (WT PCT)	< 25	25–50	> 50	LARGE STONES

TABLE 35/*Soil survey interpretations for gravel (adapted from Soil Survey Staff, 1978).*

| PROPERTY | LIMITS | | RESTRICTIVE FEATURE |
	PROBABLE SOURCE	IMPROBABLE SOURCE	
1. UNIFIED	GW, GP, GW–GM, GP–GM SW, SP, SW–SM, SP–SM		
		SW, SP, SW–SM, SP–SM	TOO SANDY
		ALL OTHER	EXCESS FINES
2. LAYER THICKNESS (IN)	> 36	< 36	THIN LAYER
3. FRACTION > 3 IN (WT PCT)	< 50	> 50	LARGE STONES

TABLE 36/*Soil survey interpretations for sand (adapted from Soil Survey Staff, 1978).*

	LIMITS		
PROPERTY	PROBABLE SOURCE	IMPROBABLE SOURCE	RESTRICTIVE FEATURE
1. UNIFIED	SW, SP, SW–SM, SP–SM GW, GP, GW–GM, GP–GM		
		GW, GP, GW–GM, GP–GM	SMALL STONES
		ALL OTHER	EXCESS FINES
2. LAYER THICKNESS (IN)	> 36	< 36	THIN LAYER
3. FRACTION > 3 IN (WT PCT)	< 50	> 50	LARGE STONES

TABLE 37/*Soil survey interpretations for roadfill (adapted from Soil Survey Staff, 1978).*

	LIMITS			RESTRICTIVE FEATURE
PROPERTY	GOOD	FAIR	POOR	
1. USDA TEXTURE	—	—	ICE	PERMAFROST
2. DEPTH TO BEDROCK (IN)	> 60	40–60	< 40	AREA RECLAIM
3. AASHTO GROUP INDEX NUMBER	0–4	5–8	> 8	LOW STRENGTH
4. AASHTO	—	A-4	A-5, A-6, A-7, A-8	LOW STRENGTH
5. LAYER THICKNESS (IN)	> 60	30–60	< 30	THIN LAYER
6. FRACTION > 3 IN (WT PCT)	< 25	25–50	> 50	LARGE STONES
7. DEPTH TO HIGH WATER TABLE (FT)	> 3	1–3	< 1	WETNESS
8. SLOPE (PCT)	0–15	15–25	> 25	SLOPE
9. SHRINK-SWELL	LOW	MODERATE	HIGH	SHRINK-SWELL

Waste disposal

PURPOSE Waste disposal is a major problem in every society, and all of us contribute to the problem. It is the purpose of this exercise to illustrate some of the soil characteristics that are important for disposal of wastes. Application of wastes to soils is becoming increasingly popular (Loehr, 1977) by applying liquids or solids onto or within the soils. Solid wastes are often buried, and liquids can be ponded, irrigated, or injected into soils. Many waste products require several treatments where liquids are separated from solids and the different components are applied to soils under contrasting situations. Study pages 95–97 of the textbook in preparation for this exercise.

SOIL PROPERTIES On a small scale, household absorption fields provide effective filtration of effluent from tile lines after solids have settled out in septic tanks. Table 20 gives soil properties (in computer format of the National Soils Handbook) important in determining limitations on performance of these systems (Figure 41). For effluent seepage, soils should be permeable without high water tables or flooding, on gentle slopes, and deep to bedrock. In your project area, locate the best soils with such ideal properties where homesites would be feasible. Absorption fields are not suitable for dense housing with large waste volumes, but function best in soils with good characteristics where the houses are isolated or not too close to one another. Location and design must be planned so that the effluent does not contaminate well water supplies. With planning, alternative designs of absorption fields can be prescribed for specific soils (Huddleston and Olson, 1967) to overcome soil limitations, including slow permeability, high water tables, shallow depth to bedrock or hardpan, slopes, flooding, and so on. The more severe the soil limitations, the greater the cost necessary to overcome the limitations.

Wastes, mostly liquid, can be treated in sewage lagoons (Table 38), which are feasible in soils with slow permeability. Slope is more limiting, because a leveled floor is necessary for oxidation, digestion, and settling of the solids. After lagoons are used for an extended period, they must be drained and the settled solids must be removed. Often, the solids are placed in landfills or spread on soils. In your project area of study, select places with slowly permeable soils that might be suitable for lagoons. These lagoons are often effective in warm or mild climates where housing is too dense for septic tank absorption fields. Lagoons are also often used to treat certain kinds of agricultural and industrial wastes. Sometimes it is feasible to further treat liquids from sewage lagoons and oxidation ponds by irrigation onto suitable soils (Table 26).

Landfills are used where garbage and other solid wastes are buried in soils. Table 39 lists soil criteria (in computer format of the National Soils Handbook) of slight, moderate, and severe limitations for waste burial in trenches 15 feet or more deep. Of course, soil surveys and engineering examinations in greater detail than usual are necessary before excavations start in a landfill. In the past, garbage has been dumped in swamps (Figure 42) and many groundwater and effluent problems have resulted. Where soils have rapid permeability, bentonite clays (Figure 43) imported from Wyoming may be compacted into an impermeable 4-inch layer before the waste is buried. Then the effluent is collected and treated, and the potential for groundwater pollution is much less.

In your study area or project area, select the best place for a large landfill. Criteria in Table 40 can be used where cover is to be placed over the natural soils without excavation. Table 41 gives ratings for soil materials to be used for daily cover for the wastes. In planning and designing your landfill, remember that hauling distance is a critical factor both for the waste and for the cover. Landfill problems are increasing in magnitude, and will become more serious in the future as hazardous effluents begin to move downslope from inadequately designed landfills where dangerous wastes have been buried in the past.

PLANNING In design of new communities, industrial plants, and other facilities, soil maps can be used very effectively to help plan the total developments in harmony with the environment. Thus, septic tank absorption fields can be used for widely spaced homes in the rural and suburban areas, lagoons can be used for denser community and industrial wastes largely liquid, and landfills can be used for the solid waste burial. Matching the location and design of the waste disposal facility to the soil characteristics is a critical factor.

With better management, dangerous toxic wastes should be separated and specially treated. Biodegradable wastes can be beneficial when properly applied to certain soils (Table 42). Even siting of nuclear power plants (Figure 44) should consider soil conditions for isolation of the facility, waste burial, and design of roads, foundations, and so on. Nuclear fallout is adsorbed on soil particles depending on clay content and other soil properties. Radioactivity of soils has been used to calculate erosion losses and deposition from different areas. In the past, many industrial developments were put on poor soils (Figure 45) because the land was cheap, but in the future information on soils

and the environment must be much more carefully considered to avoid repeating past mistakes. Many of our waste disposal problems can be solved and mistakes avoided by increased use of soil maps together with other environmental information.

REFERENCES

Huddleston, J. H. and G. W. Olson. 1967. Soil survey interpretation for subsurface sewage disposal. Soil Science 104:401–409.

Loehr, R. C. (Editor). 1977. Land as a waste management alternative: Proceedings of the 1976 Cornell Agricultural Waste Management Conference, Ann Arbor, MI. 811 pages.

Soil Survey Staff. 1978. National soils handbook. Part II, Section 400, Application of soil survey information, Procedures guide. Soil Conservation Service, U.S. Dept. of Agriculture, Washington, DC. Reproduced as Cornell Agronomy Mimeo 82-15, Cornell University, Ithaca, NY. About 100 pages.

FIGURE 41/*Tile lines (on specified grade) for a septic tank absorption field.*

FIGURE 42/*Garbage dump in swampy soils in Syracuse, New York. Groundwater is contaminated and water supplies are polluted where landfills are not properly designed to fit the soil conditions.*

FIGURE 43/*Compaction of bentonite layer at the bottom of a landfill under construction near Oswego, New York.*

FIGURE 44/*Nuclear power plant under construction near Oswego, New York. The soil characteristics were important in the initial siting of the plant, and in the design of certain aspects of the facilities.*

FIGURE 45/*Industrial plant and wetland pollution along the Carolinas Coastal Plain.*

TABLE 38/*Soil survey interpretations for sewage lagoons (adapted from Soil Survey Staff, 1978).*

PROPERTY	LIMITS			RESTRICTIVE FEATURE
	SLIGHT	MODERATE	SEVERE	
1. USDA TEXTURE	—	—	ICE	PERMAFROST
2. PERMEABILITY (IN/HR) (12–60 IN)	< 0.6	0.6–2.0	> 2.0	SEEPAGE
3. DEPTH TO BEDROCK (IN)	> 60	40–60	< 40	DEPTH TO ROCK
4. DEPTH TO CEMENTED PAN (IN)	> 60	40–60	< 40	CEMENTED PAN
5. FLOODING	NONE, PROTECTED	—	RARE, COMMON	FLOODS
6. SLOPE (PCT)	0–2	2–7	> 7	SLOPE
7. UNIFIED	—	OL, OH	Pt	EXCESS HUMUS
8. DEPTH TO HIGH WATER TABLE (FT)	—	—	+	PONDING
	> 5	3.5–5	0–3.5	WETNESS
9. FRACTION > 3 IN (WT PCT)	< 20	20–35	> 35	LARGE STONES

TABLE 39/*Soil survey interpretations for sanitary landfill in trenches (adapted from Soil Survey Staff, 1978).*

PROPERTY	LIMITS			RESTRICTIVE FEATURE
	SLIGHT	MODERATE	SEVERE	
1. USDA TEXTURE	—	—	ICE	PERMAFROST
2. FLOODING	NONE, PROTECTED	RARE	COMMON	FLOODS
3. DEPTH TO BEDROCK (IN)	—	—	< 72	DEPTH TO ROCK
4. DEPTH TO CEMENTED PAN (IN)				CEMENTED PAN
THICK	—	—	< 72	
THIN	—	< 72	—	
5. PERMEABILITY (IN/HR) (BOTTOM LAYER)	—	—	> 2.0	SEEPAGE
6. DEPTH TO HIGH WATER TABLE (FT):				
	—	—	+	PONDING
APPARENT	—	—	0–6	WETNESS
PERCHED	> 4	2–4	0–2	WETNESS
7. SLOPE (PCT)	0–8	8–15	> 15	SLOPE
8. USDA TEXTURE	—	CL, SC, SICL	SIC, C	TOO CLAYEY
9. USDA TEXTURE	—	LCOS, LS, LFS, LVFS	COS, S, FS, VFS, SG	TOO SANDY
10. UNIFIED	—	—	OL, OH, PT	EXCESS HUMUS
11. FRACTION > 3 IN (WT PCT)	< 20	20–35	> 35	LARGE STONES
12. SODIUM ADSORPTION RATIO (GREAT GROUP)	—	—	> 12 (NATRIC, HALIC)	EXCESS SODIUM
13. SOIL REACTION (pH)	—	—	< 3.6	TOO ACID
14. SALINITY (MMHOS/CM)	—	—	> 16	EXCESS SALT

TABLE 40/*Soil survey interpretations for area sanitary landfill (adapted from Soil Survey Staff, 1978).*

PROPERTY	LIMITS			RESTRICTIVE FEATURE
	SLIGHT	MODERATE	SEVERE	
1. USDA TEXTURE	—	—	ICE	PERMAFROST
2. FLOODING	NONE, PROTECTED	RARE	COMMON	FLOODS
3. DEPTH TO BEDROCK (IN)	> 60	40–60	< 40	DEPTH TO ROCK
4. DEPTH TO CEMENTED PAN (IN)	> 60	40–60	< 40	CEMENTED PAN
5. PERMEABILITY (IN/HR) (10–40 IN)	—	—	> 2.0	SEEPAGE
6. DEPTH TO HIGH WATER TABLE (FT)				
	—	—	+	PONDING
APPARENT	> 5	3.5–5	0–3.5	WETNESS
PERCHED	> 3	1.5–3	0–1.5	WETNESS
7. SLOPE (PCT)	0–8	8–15	> 15	SLOPE

TABLE 41/Soil survey interpretations for daily cover for landfill (adapted from Soil Survey Staff, 1978).

| PROPERTY | LIMITS | | | RESTRICTIVE FEATURE |
	GOOD	FAIR	POOR	
1. USDA TEXTURE	—	—	ICE	PERMAFROST
2. DEPTH TO BEDROCK (IN)	> 60	40-60	< 40	AREA RECLAIM
3. DEPTH TO CEMENTED PAN (IN)	> 60	40-60	< 40	AREA RECLAIM
4. UNIFIED	—	—	SP, SW, SP-SM, SW-SM, GP, GW, GP-GM, GW-GM	SEEPAGE
5. USDA TEXTURE	—	CL, SICL, SC	SIC, C	TOO CLAYEY
6. USDA TEXTURE	—	LCOS, LS, LFS, VFS	S, FS, COS, SG	TOO SANDY
7. UNIFIED	—	—	OL, OH, CH, MH	HARD TO PACK
8. COARSE FRAGMENTS (PCT)	< 25	25-50	> 50	SMALL STONES
9. FRACTION > 3 IN (WT PCT)	< 25	25-50	> 50	LARGE STONES
10. SLOPE (PCT)	0-8	8-15	> 15	SLOPE
11. DEPTH TO HIGH WATER TABLE (FT)	—	—	+	PONDING
	> 3.5	1.5-3.5	< 1.5	WETNESS
12. UNIFIED	—	—	PT	EXCESS HUMUS
13. LAYER THICKNESS (IN)	> 60	40-60	< 40	THIN LAYER
14. SOIL REACTION (pH)	—	—	< 3.6	TOO ACID
15. SALINITY (MMHOS/CM)	—	—	> 16	EXCESS SALT
16. SODIUM ADSORPTION RATIO (GREAT GROUP)	—	—	> 12 (HALIC, NATRIC)	EXCESS SODIUM

TABLE 42/Ratings of limitations of soils for application of biodegradable solids and liquids (adapted from Loehr, 1977).

| Item Affecting Use | Soil-Limitation Rating[a] | | |
	Slight	Moderate	Severe
Permeability of the most restricting layer above 60 in.	0.6-6.0 in./hr	6-20 and 0.2-0.6 in./hr	> 20 and < 0.2 in./hr
Soil drainage class	Well drained and moderately well drained	Somewhat excessively drained and somewhat poorly drained	Excessively drained, poorly drained and very poorly drained
Runoff	Ponded, very slow, and slow	Medium	Rapid and very rapid
Flooding	None	None for solids, only during nongrowing season allowable for liquids	Flooded during growing season (liquids) or any-time (solids)
Available water capacity from 0 to 60 in. or to a root-limiting layer	> 8 in. (humid regions) > 3 in. (arid regions)	3-8 in. (humid regions) Moderate class not used in arid regions	< 3 in. (humid regions) < 3 in. (arid regions)

[a]Moderate and severe limitations do not apply for soils with permeability < 0.6 in./hr: (1) for solid wastes unless the waste is plowed or injected into the layers having this permeability or evapotranspiration is less than water added by precipitation and irrigation, and (2) for liquid wastes if layers having that permeability are below the rooting depth and evapotranspiration exceeds water added by precipitation and irrigation.

Agricultural land classification

PURPOSE Land classifications are widely used to evaluate potentials and problems in soil management. This exercise is designed to acquaint the student with some of the classification schemes currently in extensive use, and to provide some experience in placing soil map units into some of the various categories. The student should also appreciate that different land classifications are designed for different purposes, and no one system should be expected to serve all possible needs. Chapter 7 (pages 98–104) in the textbook, as well as the references, should be studied in preparation for this exercise.

PROCEDURE From detailed criteria in the references, classify the soils of your project area in several different land classification systems. Soils Bulletin 22 (FAO, 1974) gives a summary of the most commonly used land classifications in the world, and Olson (1974) has reviewed the systems in wide use in English-speaking countries. Compare and contrast the placement of your soils in the different systems. The land capability classification is designed primarily for erosion control, the irrigation suitability classification is for irrigation development, the Storie index was intended to provide a numerical rating of soils, the prime farmland definitions help to locate those areas that should be preserved for agriculture, the preferential tax assessment is designed to provide a fair land tax, the FAO system was to inventory the productive potential of soils of the world, and formulas for calculating a productivity index have factors based on the sufficiency of soil properties to meet the crop needs. Land classifications can be designed for farmland, pasture, range, forestry, or for other purposes. Briefly, some criteria and descriptions are given here to introduce the student to some of the concepts, and to lead the student into some of the details of some of the land classifications.

CLASSIFICATIONS Many land classifications have been devised by different authors and agencies. The monograph by Beatty et al. (1979) reviews some of the different perspectives. Table 43 lists soil properties and data important in some

land classification systems. The Storie index system (Storie, 1964) was designed initially for use in California. The U.S. Department of Agriculture (USDA) system (Klingebiel and Montgomery, 1973) was made primarily for erosion control on U.S. soils. The Canadian system (Anonymous, 1972) was primarily a system for inventory and evaluation. Beek and Bennema (1972) formulated their classification for agricultural land-use planning in The Netherlands. Obeng (1968) devised his system for soils of Ghana under practices of mechanized and hand cultivation for crop and livestock production. The paper by Bartelli (1968) is oriented to potential farming lands in the coastal plain of the southeastern United States. Each system is useful in its particular environment and circumstance, but none are appropriate for all uses at all times. Students in their project efforts should strive to achieve the most useful system to meet the needs of their specific project area, by adapting and revising the existing systems or formulating new ones.

LAND CAPABILITY One of the most widely used systems of land classification is that of the Soil Conservation Service of the U.S. Department of Agriculture, commonly called land capability classification. It is summarized in a bulletin by Klingebiel and Montgomery (1973). A bulletin and set of 50 slides introduce the concepts of the classification in a form readily understood by laypersons (SCS, 1969).

The land capability classification is based on the detailed soil survey, generally published at scales of about 1:20,000 or 1:15,840 in the United States (see Figures 2 and 3). The classification consists essentially of grouping the various soil map units "primarily on the basis of their capability to produce common cultivated crops and pasture plants without deterioration over a long period of time" (Klingebiel and Montgomery, 1973). The soil map unit is defined as the "portion of the landscape that has similar characteristics and qualities and whose limits are fixed by precise definitions"; the soil map unit is the entity about which the greatest number of precise statements and predictions can be made.

Capability units, into which soil map units are grouped, have similar potentials and continuing limitations or hazards. Soil map units put into a capability unit are sufficiently uniform to produce similar kinds of cultivated crops and pasture plants with similar management practices, require similar conservation treatment and management under the same kind and condition of vegetative cover, or have comparable potential productivity. Use of capability units condenses and simplifies soil map unit information for planning the use and management of individual areas of land, even as small as several acres in size. A capability unit is designated by a symbol such as IIIe2. The Roman numeral designates the capability class of lands that have the same relative degree of hazard or limitation; the risks of soil damage or limitation in use became progressively greater from Class I to Class VIII. The lowercase letters designate subclasses that have the same major conservation problem: e (erosion and runoff), w (excess water), s (root zone limitations), and c (climatic limitations). The Arabic numbers indicate the capability unit within each capability class and subclass.

Land capability classes introduce the map user to the more detailed information on the soil map; a map of the classes shows the location, amount, and general suitability of the soils for agricultural purposes. The land capability subclasses provide information about the kind of conservation problem or limitation involved in land use. When both the land capability class and subclass are used together, they provide the map reader with general information about the degree of limitations and kind of problem involved for broad program planning, studies of conservation needs, and similar purposes. The capability unit is a grouping of soil areas that are enough alike to be suited to the same crops and pasture plants, to require comparable management, and to be somewhat similar in productivity and in other responses to management.

Figure 46 illustrates landscape segmentation into capability classes in California. Table 44 shows the criteria on which the decisions were made to place soil map units into capability classes. Each region in the United States provides such guidelines to states and to Soil and Water Conservation Districts for placement of soils into capability classes in a standardized systematic format. Figure 47 is the guide used for assigning land classes and subclasses to soils in the northeastern United States and New York State. Students and others can follow the "guide" or other similar guides to assign subclass designations to specific soil map units of their project area.

Hord silt loam soils on 1 to 3 percent slopes in south central Nebraska, for example, are classified IIc1 (Figure 48). The "c" designation indicates that the soils have a climatic limitation due to the dry climate, but the productivity for corn is excellent when the areas are irrigated. In contrast, Sharkey clay mapped near New Orleans, Louisiana (Figure 49), is classified IIIw—indicating wetness. Where drained, yields of Sharkey clay and similar soils are much improved for crops such as sugarcane.

STORIE INDEX Land classification by quantitative productivity indexes has been used in many places. The Storie index, developed at the University of California, illustrates principles of application of productivity indexes in land classification that have been relatively widely applied (Edwards et al., 1970). The Storie index method has undergone a number of revisions over the years, as more data have been gathered and more experience has been obtained in using it; some of the principles and revisions are published in bulletins of the University of California and in papers in the *Proceedings* of the Soil Science Society of America. A good example of application of the index concept is given in the detailed land classification of the island of Oahu in the Hawaiian Islands (Nelson et al., 1963).

Master ratings developed for Oahu lands were based on general character of the soil profile, texture of the surface soil, slope of the land, climate, and other physical conditions affecting use of the land; ranges in percentage values were selected for the various factors that were appropriate for the local conditions. The land productivity index can be stated as follows:

$$\text{land productivity index} = A \times B \times C \times X \times Y$$

where

A = percentage rating for the general character of the soil profile

B = percentage rating for the texture of the surface horizon

C = percentage rating for the slope of the land

X = percentage rating for site conditions other than those covered in factors A, B, and C (e.g., salinity, soil reaction, freedom from damaging winds)

Y = percentage rating for rainfall

The land productivity index is obtained by multiplying a series of percentage ratings. Percentage ratings are converted to decimal equivalents for use in the formula, and the resulting product is reconverted to a percentage basis. The percentage rating for each factor (A, B, C, X, and Y) increases as the favorableness of that factor increases. As the land productivity index approaches 100 percent, the agricultural quality of the land increases. Less productive land types have indexes with lower values. If even one factor alone has a low percentage rating, that factor can substantially reduce the level of the land productivity index.

The ratings for factor A of the productivity index take drainage and thickness of the soil profile into consideration (Nelson et al., 1963). Several degrees of drainage are recognized. Deep and shallow soils have been separated; the nature of surface soil and subsoil is important here. Parent material and degree of soil development are critical determinants of nutrient-supplying capacity, volume of soil available for root development, and physical properties such as soil structure, aeration, and moisture-supplying capacity.

Percentage ratings for factor B (surface texture) are expressed in terms of textural classes that portray the relative proportions of sand, silt, and clay. Special categories are provided for stony lands, rocky lands, and pahoehoe. Major soil types of a series are generally consolidated within the textural groups. Texture of a soil is clearly associated with physical properties such as water-holding capacity and the ability of the soil to supply moisture to plants. In the clay fraction, many important biochemical and physiochemical reactions take place, including organic absorption by clays, ion exchange, and nutrient fixation.

Ratings for factor C indicate the general slope of land. The slope classes were developed to differentiate ease of irrigation and use of mechanical equipment, susceptibility to erosion, amount of surface runoff, and suitability for commercial forest production. In general, slopes greater than 35 percent are not suitable for production of cultivated crops, and slopes steeper than 80 percent are impractical for commercial forest production.

Factor X is a composite component because it is a product of percentage ratings for several factors, including soil reaction, salinity, nutrient status, erosion, and wind damage hazards (Nelson et al., 1963). On Oahu, the availability of most nutrient elements is favored by a neutral or slightly acid soil

reaction. Some Oahu soils have high NaCl, particularly in upland areas that are being irrigated with brackish water, and in lowland coastal plains subject to capillary action or spray effects from seawater. Soil tests are made for phosphorus, potassium, and calcium in Oahu soils; most soils are uniformly low in nitrogen; in a few places iron, zinc, manganese, boron, magnesium, and molybdenum may be a problem. Erosion limits potential productivity by reduction of the volume of soil available for root development, loss of organic matter, loss of nutrients, and removal of the soil having the most desirable physical condition for plant growth. The windward side of Oahu is particularly subject to wind damage, which may reduce yields by causing excessive evapotranspiration and physical injury to the plants.

Factor Y considers rainfall, and indirectly also temperature, sunlight, and cloud cover. A percentage rating of 100 is given to irrigated lands, because moisture levels can be maintained at optimum levels on these lands.

To illustrate use of the revised Storie index, the process of deriving the index value (Table 45) for Land Type 10 is as follows (Nelson et al., 1963):

land productivity index = $A \times B \times C \times X \times Y$

where

A = 97 percent because the soil is deep, well drained, and has developed in uplands or old alluvium on upland terraces

B = 90 percent because the texture of the soils is silty clay loam

C = 100 percent because the slope of the land is in the range 0 to 10 percent

X = 74 percent (Table 45) because the surface soil reaction is medium acid to slightly acid (98), the fertility level is poor (82), a moderate level of erosion has occurred (92), and wind damage and salinity pose no problems (100); the product of the percentage ratings for these factors is $0.98 \times 0.82 \times 0.92 \times 1.00 \times 1.00 = 0.74 = 74$ percent

Y = 92 percent because the annual rainfall ranges from 30 to 80 inches

Land Type 10 includes soils that are nonstony, moderately deep, well drained, and moderately fine textured, having dark, reddish-brown surface soil over red subsoil. The soils have developed in uplands or upland terrace positions in old alluvium or volcanic ash. Soil reaction is medium to slightly acid. Slopes range from 0 to 10 percent. Soil series included are Kolekole and Manana. Lands are easily worked. Special problems are high erodibility, low fertility due to intense weathering, a compact subsoil that restricts root growth to the upper horizons, and a fluffy, droughty layer above the pan. Median annual rainfall varies from 30 to 80 inches. Elevations range from 500 to 1,200 feet.

Decimal equivalents of percentage ratings for individual factors were multiplied in sequence ($A \times B \times C \times X \times Y$). The resulting product upon conversion to a percentage basis constitutes the land productivity index (Nelson et al., 1963):

$$A \times B \times C \times X \times Y = \text{land productivity index}$$
$$0.97 \times 0.90 \times 1.00 \times 0.74 \times 0.92 = 0.59$$
$$= 59 \text{ percent}$$

Master productivity ratings are assigned when productivity of all the land types of Oahu have been evaluated. Considerable amounts of research and experimentation, of course, go into the total process of determining what are the best values for each of the factors and what is the best management to recommend for each land type. Yields have been specified, for example, for lands used for growing pineapples, sugarcane, vegetables, alfalfa, pasture, oranges, papayas, bananas, and trees. This type of land classification has not only been of great value in managing the best lands for the highest yields, but it has also been used for zoning the best agricultural lands for preservation from urban encroachment—a serious agricultural problem at the present time in Hawaii.

Students should experiment with the assigning of index numbers to soil properties and calculations of overall ratings for different soils. Generally, index numbers should be as directly related as possible to real numbers (e.g., pH, base saturation, percent clay, etc.). Arbitrary numbers have less validity and are more subject to question in the statistical analyses. A good technique is to compare properties of a specific soil to properties of the best soil for a specific use, and then assign percentage ratings accordingly. A score of 100 can be assigned to the best soil in a survey area, or the best soil in a state, depending on the size of the universe for which the soil rating (land classification) system is designed to serve. All other soils then have lesser numbers as a percentage (e.g., yield) as compared with the best soil.

SUFFICIENCY CONCEPT Pierce et al. (1982) have adopted a numerical index method to quantify soil productivity based on the assumption that soil is a major determinant of crop yield due to the environment it provides for root growth. Soil parameters were evaluated in terms of root response and each soil layer was weighted according to an ideal rooting distribution. Response and each soil parameter were normalized to range from 0.0 to 1.0. Data available from the SCS SOILS 5 data base were used (see Tables 16 to 18). The model used to evaluate productivity in this approach takes the following form:

$$PI = \sum_{i=1}^{r} (A_i \times C_i \times D_i \times E_i \times WF)$$

where

A_i = sufficiency of available water capacity (adjusted for coarse fragments and salts)
C_i = sufficiency of bulk density (adjusted for permeability)
D_i = sufficiency of pH
E_i = sufficiency of electrical conductivity
WF = weighting factor
PI = productivity of the soil environment
r = number of horizons in the depth of rooting under ideal conditions

Students and others with computer facilities should adopt this formula to fit their soils environment for calculating a productivity index for each of the different project area soils. The sufficiency concept can be extremely valuable in comparing a variety of different soils, especially as the soils are influenced by management practices affecting drainage and erosion and other soil improvements and degradations (Pierce et al., 1982).

IRRIGATION SUITABILITY The land classification system of the Bureau of Reclamation of the U.S. Department of the Interior has been used or adapted in many countries for irrigation project areas. Students should become familiar with it, especially if their project area is likely to be irrigated or if they plan to work in arid regions in the future. A good general summary of the land classification is given in the monograph on irrigation of agricultural lands (Hagen et al., 1967). Feasibility is determined by payment capacity:

$$Y = -a + bX_1 - cX_2 - dX_3$$

where

$$Y = \text{payment capacity (dollars)}$$
$$X_1 = \text{productivity rating (percent)}$$
$$X_2 = \text{land development cost (dollars)}$$
$$X_3 = \text{farm drainage cost (dollars)}$$
$$a, b, c, d = \text{constants derived from farm budget analyses}$$

In a given project area:

$$E = (S, T, D)$$

where

$$E = \text{economic parameter}$$
$$S = \text{soil characteristics}$$
$$D = \text{drainage characteristics}$$
$$T = \text{topographic characteristics}$$

The factors S, T, and D are considered everywhere, but the individual characteristics of each, such as texture, structure, horizon arrangement, depth, salinity and alkalinity of S, macrorelief and microrelief of T, and surface and subsurface drainage of D, are selected on the basis of relevance to prediction of E at the given time and place. For land classification purposes, the quality of land for irrigation use can then be indicated by land classes that represent specified meaningful ranges in the value of E (Hagen et al., 1967).

The definitions of land classes (Table 46) are as follows:

Class 1 Arable lands of Class 1 are those particularly suitable for irrigation farming, being capable of producing sustained and relatively high yields of a wide range of climatically adapted crops at reasonable cost. They are smooth lying with gentle slopes. The soils are deep and of medium to fairly fine texture with mellow open structure allowing easy penetration of roots, air, and water and having free drainage, yet good available moisture capacity. These soils are free of harmful accumulations of soluble salts or can be readily reclaimed. Both soil and topographic conditions are such that no specific farm drainage requirements are anticipated, minimum erosion will result from irrigation, and land development can be accomplished at fairly low cost. These lands potentially have a relatively high payment capacity.

Class 2 Arable lands of Class 2 comprise those that are moderately suitable for irrigation farming, being measurably lower than Class 1 lands in productive capacity, adapted to a somewhat narrower range of crops, more expensive to prepare for irrigation, or more costly to farm. They are not as desirable or of such high value as lands of Class 1 because of certain correctable or noncorrectable limitations. They may have a lower available moisture capacity, as indicated by coarse texture or limited soil depth; they may be only slowly permeable to water because of clay layers or compaction in the subsoil; or they also may be moderately saline, which may limit productivity or involve moderate costs for leaching. Topographic limitations include: uneven surface requiring moderate costs for leveling, short slopes requiring shorter length of runs, or steeper slopes necessitating special care and greater costs to irrigate and prevent erosion. Farm drainage may be required at a moderate cost, or loose rock or woody vegetation may have to be removed from the surface. Any one of the limitations may be sufficient to reduce the lands from Class 1 to Class 2, but frequently a combination of two or more of them is operating. The Class 2 lands have intermediate payment capacity.

Class 3 Arable lands of Class 3 are somewhat suitable, but approaching marginality, for irrigation development and are of distinctly restricted suitability because of more extreme deficiencies in the soil, topographic, or drainage characteristics than those described for Class 2 lands. They may have good topography, but the inferior soils restrict crop adaptability, require larger amounts of irrigation water or special irrigation practices, and demand greater fertilization or more intensive soil improvement. The topography may be uneven, saline concentration moderate to high, and drainage restricted but correctable at relatively high costs. Generally, a greater risk may be involved in farming Class 3 lands than the better classes of land, but under proper management they are expected to pay adequately.

Class 4 Limited arable or special-use lands of Class 4 are included in this class only after economic and engineering studies have shown them to be tillable. They may have an excessive specific deficiency or deficiencies that can be corrected at high cost, but are suitable for irrigation because of present or contemplated intensive cropping, such as for vegetables and fruits. They may have one or more excessive, noncorrectable deficiencies that limit their use to meadow, pasture, orchard, or other relatively permanent crops, but still be capable of supporting a farm family and meeting water charges if operated in units of adequate size or in association with better lands. The deficiency may be inadequate drainage, salt content that requires extensive leaching, unfavorable position allowing periodic flooding or making water distribution and removal difficult, rough topography, excessive quantities of loose rock on the surface or in the flow zone, or cover such as timber. The magnitude of the correctable deficiency is sufficient to require outlays of capital in excess of those permissible for Class 3 but in amounts shown to be feasible because of the specific utility anticipated. Sub-classes other than those devoted to a special crop use may be included in this class, such as those for subirrigation and sprinkler irrigation which meet general arability requirements. Also recognized in Class 4 are suburban lands that do not meet general arability requirements. Such lands can pay water charges as a result of income derived either from the suburban lands and other souces or from other sources alone. The Class 4 lands may have a range in payment capacity greater than that of the associated arable lands.

Class 5 Nonarable lands of Class 5 are nonarable under existing conditions but have a potential sufficient to warrant tentative segregation for special study before the classification is completed. They may also be lands in existing projects whose arability is dependent on scheduled project con-struction or land improvements. They may have a specific soil deficiency such as excessive salinity, very uneven topography, inadequate drainage, or excessive rocks or tree cover. In the first instance, the land deficiency or deficiencies are of such kind and magnitude that special agronomic, economic, or engineering studies are required to provide adequate information (such as for extent and location of farm and project drains, or probable payment capacity under the anticipated land use) for completing the classification of the lands. The designation of Class 5 is tentative and must be changed to the proper arable class or Class 6 when the land classification is completed. In the second instance, the effect of the deficiency or the outlay necessary for improvement is known, but the lands are suspended from an arable class until the scheduled date of completion of project facilities

and land development, such as project and farm drains. In all instances, Class 5 lands are segregated only when the conditions in the area require consideration of such lands for competent appraisal of the project possibilities, such as when an abundant supply of water or shortage of better lands occurs, or when problems related to land development, rehabilitation, and resettlement are involved.

Class 6 Nonarable lands of Class 6 include: those considered nonarable under the existing project or project plan because of failure to meet the mini-mum requirements for the other classes of land; arable areas definitely not susceptible to delivery of irrigation water or to provision of project drain-age; and Classes 4 and 5 land, when the extent of such lands or the detail of the particular investiga-tion does not warrant their segregation. Generally, Class 6 lands comprise steep, rough, broken, or badly eroded lands; lands with soils of very coarse or fine texture; lands with shallow soils over gravel, shale, sandstone, or hardpan; or lands that have inadequate drainage and high concentrations of soluble salts or sodium. With some exceptions, Class 6 lands do not have sufficient payment capacity to warrant consideration for irrigation.

Detailed land classification for irrigation (Hagen et al., 1967) is generally done at a map scale of 1:4,800 (400 feet to the inch) to provide adequate information as to the extent and char-acter of the various soils in each 40-acre tract. A smaller scale, not less than 1:12,000, may be used on fully developed areas or on highly uniform new land areas where no specific problems are associ-ated with soils, topography, or drainage, and none are anticipated. Base maps at scales of 1:24,000 are considered only for reconnaissance studies, and are used for preliminary evaluations and for drain-age basin studies (e.g., runoff, conservation) of areas not to be irrigated, but within the general project area.

At the large scale at which soils are mapped in irrigation projects, a great deal of soil information can be put on the map by use of letters, numbers, and symbols. Table 47 illustrates examples of some standard mapping symbols and components used for land classification surveys. A numeral—1, 2, 3, 4, or 6—designation is used for the appraisals that relate to the appropriate land class level. An estimation of farm water requirements is made using the letters A, B, and C to indicate low, medium, and high. The letters X, Y, and Z indicate good, restricted, and poor drainability conditions in the 5- to 10-foot zone. Additional informative appraisals are made using g for slope, u for undulations, f for flooding, k for shallow depth to sand, gravel, cobbles, and so on. Areas of different land classes on maps are colored yellow (Class 1), green (Class 2), blue (Class 3), brown (Class 4), and pink (Class 5) to make the maps more understandable to economists, hydrologists, and engineers concerned with the development of the plan (nonarable land is not colored). Results of soil profile examinations and laboratory analyses are also put on the map where appropriate. Field surveys are generally supplemented with extensive laboratory analyses, greenhouse studies, and field experimental plot data to obtain as much information as is needed before the irrigation project is implemented. Voluminous reports summarizing the data accompany the maps at the various scales.

PRIME FARMLAND To identify areas of the best soils that should be considered for preservation for future food production, definitions of "prime farmlands" were established (Beatty et al., 1979).

1. The soils have one of the following:
   a. Aquic, udic, ustic, or xeric moisture regimes and sufficient available water capacity within a depth of 1 m (40 in.), or in the root zone if the root zone is less than 1 m deep, to produce the commonly grown crops in 7 or more years out of 10; or
   b. Xeric or ustic moisture regimes in which the available water capacity is limited, but the area has a developed irrigation water supply that is dependable (a dependable water supply is one in which enough water is available for irrigation in 8 out of 10 years for the crops commonly grown) and of adequate quality; or
   c. Aridic or torric moisture regimes and the area has a developed irrigation water supply that is dependable and of adequate quality; and

2. The soils have a soil temperature regime that is frigid, mesic, thermic, or hyperthermic (pergelic and cryic regimes are excluded). These are soils that, at a depth of 50 cm (20 in.), have a mean annual temperature higher than 0°C (32°F). In addition, the mean summer temperature at this depth in soils with an O horizon is

higher than 8°C (47°F); in soils that have no 0 horizon the mean summer temperature is higher than 15°C (59°F); and

3. The soils have a pH between 4.5 and 8.4 in all horizons within a depth of 1 m (40 in.) or in the root zone if the root zone is less than 1 m (40 in.) deep. This range of pH is favorable for growing a wide variety of crops without adding large amounts of amendments; and

4. The soils either have no water table or have a water table that is maintained at a sufficient depth during the cropping season to allow food, feed, fiber, forage, and oilseed crops common to the area to be grown; and

5. The soils can be managed so that in all horizons within a depth of 1 m (40 in.), or in the root zone if the root zone is less than 1 m (40 in.) deep, during part of each year the conductivity of saturation extract is less than 4 mmhos/cm and the exchangeable sodium percentage (ESP) is less than 15; and

6. The soils are not flooded frequently during the growing season (less often than once in 2 years); and

7. The soils have a product of K (erodibility factor) X percent slope which is less than 2.0 and a wind erosion product of I (soil erodibility) X C (climate factor) not exceeding 60. That is, prime farmland does not include soils having a serious erosion hazard; and

8. The soils have a permeability rate of at least 0.15 cm (0.06 in.) per hour in the upper 50 cm (20 in.) and the mean annual soil temperature at a depth of 50 cm (20 in.) is less than 15°C (59°F); the permeability rate is not a limiting factor if the mean annual soil temperature is 15°C (50°F) or higher; and

9. Less than 10% of the surface layer (upper 15 cm) in these soils consists of rock fragments coarser than 7.6 cm (3 in.). These soils present no particular difficulty in cultivating with large equipment.

In general, this prime farmland definition (Beatty et al., 1979) embraces all land in Capability Class I, most of Class II, and Subclass IIIw that has an adequate water management system. These criteria for prime farmland exclude land in Subclass IIw if the water table is not maintained at a sufficient depth to allow growing crops common to the area, those soils where there is not enough moisture to permit annual cropping (summer fallow areas), and those soils that have a serious wind erosion hazard. Although it is tempting simply to list the capability classes (see textbook pages 98–100) and subclasses that qualify as prime farmland, these classes do not necessarily reflect the productive capacity of soils. Students should acquire a recognition of the reasons for the differences in the classification systems, and know the soil characteristics that are responsible for the overlap and nonconformities in the different categories.

ASSESSMENT FOR TAXATION Detailed soil maps are being increasingly used for taxation, because they provide a valid, fair, unbiased appraisal of productivity potential. Table 48 summarizes some of the features of the systems used by the various states in the United States. In New York State, taxation of agricultural lands is assessed from the productivity of the soil map units. Yield estimates from the soil survey report and SCS Form 5 (see Tables 16 to 18) are converted to TDN (total digestible nutrients) for corn and hay, from the recommended rotation based on the soil loss equation. The TDN yield values are placed into "mineral soil groups" from index numbers derived from TDN ratios (maximum TDN yield/TDN soil map unit yield $\times$ 100). Table 49 is an example of one page of the computer printout of soil map units showing lime level, rotation, corn yields, hay yields, TDN yields, and the mineral soil group for each soil map unit. The total digestible nutrients are 20 percent of the yield weight of corn silage and 50 percent of the yield weight of hay.

Index yields are given for the mineral soil groups in Table 50; woodland yields are listed in board feet per acre. Table 51 gives the average soil value on which the land tax was based. From the yield data and costs, economic profiles were estimated and calculated for each mineral soil group and organic soil group and woodland group. Agricultural value per acre is adjusted each year to keep current and to meet the taxation budget goals. The value per acre is different for high- and low-lime soils in the mineral soil groups, and different between upstate New York and Long Island. Adjustments have been made to convert estimates of yields at a high level of management to average management levels.

The tax structure offers enormous potential for improving land use. Incentives can be provided to grow certain crops or to achieve high yields, and

penalties can be imposed if management is inadequate (e.g., if soil erosion is excessive). Farmers and others are convinced that such a tax is fair, and wide acceptance has been achieved. Of course, grumblings are always heard that taxes are too high —but social needs and priorities must be ultimately established in accord with the resource base and the true ability of the landowner to pay the tax. Students as potential landowners and all present landowners should be fully aware of all the implications in the use of detailed soil maps in assessment for taxation.

GERMAN EXPERIENCE Probably the most detailed land classification system in any country was legislated in Germany in 1934 as part of a tax reform. This sophisticated system classifies every hectare of nonurban land (Weiers and Reid, 1974). The land classification aspect of the tax reform was to achieve four aims:

1. A fair distribution of taxes.
2. An improved basis for the distribution of grants and loans to agriculture.
3. A basis for the food production planning, amalgamation, and drainage of farms.
4. A basis for land-use planning and the technical work of the agricultural advisory service.

The soil and land-use classification effort consisted of teams of workers digging pits and systematically making notes and maps. Every hectare was classified according to land use: (1) arable use, (2) garden and horticultural land, (3) permanent grassland pasture for grazing, (4) permanent grassland meadows, (5) small meadows mainly forested, (6) rough grazing, (7) temporary grass break on arable land, and (8) temporary arable land (arable crops sometimes grown between grass breaks). Much of German agricultural land use was mapped in the fifteenth century, so that changes in 1934 were also recorded. Soils were identified and described according to their nature and character. Yield potential was determined considering soil nature, soil conditions, topography, and climate conditions. Maps were drawn at scale 1:100 up to 1:5,000, depending on the uniformity of soils, the size of fields, and the structure of farming in the particular area.

Figure 50 outlines the scheme for the calculation of the soil–climate index. Textures classified were: (1) sand, (2) slightly loamy sand, (3) loamy sand, (4) heavy loamy sand, (5) sandy loam, (6) loam,

(7) heavy loam or clay loam, (8) clay, and (9) peat. Parent materials identified were: (1) diluvial soil on drift of glacial, periglacial, or colluvial origin, (2) loess, (3) alluvium, (4) sedimentary, and (5) stony. Soil profile conditions ranged from Class 1 with a uniform good well-drained profile with gradual boundaries to Class 7 with acid abrupt boundaries and mottles. Pasture was classified separately from arable land. The annual average temperature was used to differentiate climate regimes: (1) $> 8.0°C$, (2) 7.0 to 7.9°C, (3) 5.7 to 6.9°C, and (4) $< 5.6°C$ (mountain areas). Water conditions in the soil ranged from Class 1 (well drained) to Class 5 (wet, marshy ground).

After every hectare of soil had been identified and described, then a 100-point scale yield potential system was established. The best soil–climate conditions in 1934 were considered to be near Magdeburg, but this area is now in East Germany, so that a new base area had to be later established near Hildesheim. The valuation (Figure 51) was done by a Commission appointed by the Finance Minister, represented by the presidents of the regional tax offices. The local valuation commission consisted of several persons headed by the leader of the local tax administration office, normally assisted by a specially trained agricultural tax surveyor. One member from the farmers' side and one member of the agricultural administration sector (advisory service, agricultural schools) were also members of the commission, as well as a member from the district surveyor's office. Each local commission had at least four workers to dig soil pits from 50 to 200 cm in depth. The farmers were also legally bound to provide any assistance required by the valuation commission, and the commission was allowed to dig holes down to 2 meters in depth in every field without compensation to the farmers. Generalized yields in relation to the soil–climate index are illustrated in Table 52.

After the finalization of the soil classification, the results were made public on a parish area basis (Weiers and Reid, 1974). A certain time was allowed for the landowners or anybody else to raise objections to these classifications. Afterward, the results of the soil classification were passed on to the surveying authorities and through them recorded in a real estate register. As Figure 51 illustrates, many adjustments were made for economic conditions, management factors, and special deviations from the standard.

Every page in the land register has an equivalent page in the cadastral register (Weiers and Reid, 1974). The cadastral register is kept by the district authorities and organized in separate card index systems on a parish area basis. The descriptive part of the field and the soil is much more comprehensive compared with the location register. Here are found names and addresses of owners, the number of the land register, and the page in the land register where this particular field is mentioned in the cadastral register. The tax office of the district where the field is located, the tax office where the owner lives, and the land registry have a copy of the cadastral register. This system provides at least three institutions with common basic information about the soil and land.

The German land classification system has been used in countless ways. The yield index is widely recognized as an indicator of soil quality and the value of the farm. Table 53 illustrates some of the relationships between yield index and the price of the farm. Agricultural rents are 4 to 6 DM per hectare per year for each unit of the yield index. The standard valuation is the basis for the land tax, probate valuation, refugee tax, wealth tax, inheritance tax, agricultural levy, and so on. The soil classification and standard valuation have also been very helpful in regional planning, urban development, motorway routing, and so on. With computers, manipulations of the data could be much facilitated in the future.

The German experience with land classification, so thorough and advanced for the time, should illustrate to us that land resource management is closely linked to the ethics and morality of the whole of each society. German energies during 1935–1945 would have been much more productive if devoted toward peaceful land management and agricultural improvement than toward military expenditures and war. For the future, the German experience should provide examples of excellent resource inventory techniques but ultimate resource wastage by the society. This is not the first time that soils have been mismanaged to the detriment of a society (see pages 130–142 of the textbook). Through more emphasis on worthwhile and productive ultimate uses of soil information (soil survey interpretations), perhaps we (teachers and students and laypersons) can help to avoid repeating the mistakes of the past.

FAO AGRO-ECOLOGICAL ZONES The project of delineating "agro-ecological zones," based on the FAO World Soil Map, is described in general terms on pages 143–146 of the textbook. Many publications describe this project, and a library of relevant FAO reports can be accumulated for teaching purposes. World Soil Resources Report 48 (FAO, 1978), for example, illustrates the methodology and results for Africa. A land suitability assessment is made, based on the soils and climate, and requirements for various crops are matched to the land inventory according to different management levels. Reports on similar inventories have been published for Canada (Small, 1982; Dumanski and Stewart, 1981). Students can also use this approach on their project areas mapped at larger scales in greater detail. The FAO and Canadian reports on "agro-ecological zones" are general in nature, but the principles they illustrate can be applied successfully in more detailed studies. The matching of crop requirements to land (soil and climate) resources is a logical application of soil survey interpretations and climate inventory.

WOODLAND SUITABILITY Suitability classification of soils for woodlands is outlined on page 86 of the textbook. Each woodland soil group is identified by a three-part symbol, such as 2o1, 2w1, or 3w1. The potential productivity is indicated by the first number; 1 means very high, 2 high, 3 moderately high, 4 moderate, 5 low, and so on. Ratings are based on tree growth of indicator species as measured in the field. The second part of the woodland soil group symbol is a small Arabic letter, such as o (none or slight limitation), w (excessive wetness), d (restricted rooting depth), s (dry, unstable, abrasive sandy soil), and r (steep slope, erosion hazard, equipment limitation). The last number in the symbol differentiates groups of soils that need further subdivision because they require different management or are suited to different species of trees. Each woodland group is also rated slight, moderate, or severe for erosion hazard, equipment limitation, seedling mortality, plant competition, and windthrow hazard. Trees for planting and to favor in a stand are also listed for each soil group. Group 2o2, for example, may include deep, well-drained level medium-textured soils formed in recent alluvium with only brief seasonal flooding. Group 3w1 may include somewhat poorly drained, moderately fine-textured

soils formed in water-sorted clayey sediments. Group 5d1 may include shallow-to-bedrock medium-textured soils. Each survey area has a different sequence of soil groups. Although the best soils for crops are generally also the best soils for trees, woodlands are often productive on soils that are marginal for other purposes (Figure 52). Students should recognize the merits of considerations of alternative land uses, and reserve part of their project area for trees, where they are desirable and adapted for watershed protection and groundwater recharge as well as for production of wood products.

RANGE Soils that have the capacity to produce the same kinds, amounts, and proportions of range plants for grazing animals are grouped into range sites (Richardson et al., 1979). A range site is the product of all environmental factors responsible for its vegetation and productive capacity. Overgrazing, burning, plowing, erosion, and so on, disrupt the climax vegetation of the range site. Four range condition classes are used to indicate the degree of departure from the potential climax vegetation. A range is in excellent condition if 76 to 100 percent of the vegetation is one of the same kind as that in the climax stand. It is in good condition if the percentage is 51 and 75; in fair condition if the percentage is 26 to 50; and in poor condition if the percentage is less than 25.

Soils in the Coronado National Forest (Richardson et al., 1979), for example, have been placed into range sites some of which are named as follows:

1. Clay bottom range site, 12 to 20-inch precipitation zone.
2. Clay loam hills range site, 12 to 16-inch precipitation zone.
3. Limestone hills range site, 12 to 16-inch precipitation zone.
4. Loam bottom range site, 12 to 20-inch precipitation zone.

Five range productivity groups (Richardson et al., 1979) have been designated according to their production of range herbage per acre when the range is in good condition and in poor condition.

Group 1: 1,500 pounds or more per acre in good condition and 700 pounds in poor condition.
Group 2: 1,100 to 1,500 pounds per acre in good

condition and 500 to 600 pounds in poor condition.

Group 3: 750 to 1,100 pounds per acre in good condition and 400 pounds in poor condition.

Group 4: 250 to 750 pounds per acre in good condition and 200 pounds in poor condition.

Group 5: 250 pounds per acre or less in good condition and 100 pounds or less in poor condition.

Management is critical for range productivity. If the proper number of animals is grazed for relatively short periods of time on each site in accord with its capabilities, productivity is at a maximum. If excessive numbers of animals are grazed for too long periods on range sites, the productivity declines. The challenge is to match the soil capability to the proper number of animals grazed on each range site.

SUMMARY This exercise demonstrates differences between land classification systems, and shows that no single grouping of soils can be adequate for all purposes. Initiative and imagination are important in grouping soils for different uses, but uniform standards are also necessary when inventories must be made of the resources of a nation and when some uniformity must be applied in making recommendations for improving management of land areas. Guidelines provided in these land classification discussions and references have been widely used over large areas, and will prove to be valuable to students and others classifying soils in project areas.

REFERENCES

Anonymous. 1972. Soil capability classification for agriculture. The Canada Land Inventory, Report 2, Department of Environment, Ottawa. 16 pages.

Bartelli, L. J. 1968. Potential farming lands in the coastal plain of southeast United States. Transactions of the 9th International Congress of Soil Science 8:243–251.

Beatty, M. T., G. W. Petersen, and L. D. Swindale (Editors). 1979. Planning the uses and management of land. Monograph 21, American Society of Agronomy, Madison, WI. 1028 pages.

Beek, K. J. and J. Bennema. 1972. Land evaluation for agricultural land use planning. Agricultural University, Dept. of Soil Science and Geology, Wageningen, The Netherlands. 70 pages.

Dumanski, J. and R. B. Stewart. 1981. Crop production potentials for land evaluation in Canada. Land Resource Research Institute, Research Branch, Agriculture Canada, Ottawa. 80 pages and map and microfilm.

Edwards, R. D., D. F. Rabey, and R. W. Kover. 1970. Soil survey of Ventura area, California. University of California Agricultural Experiment Station, and Soil Conservation Service, U.S. Dept. of Agriculture, U.S. Government Printing Office, Washington, DC. 151 pages and 50 soil map sheets.

FAO. 1974. Approaches to land classification. Soils Bulletin 22, Food and Agriculture Organization of the United Nations, Rome, Italy. 120 pages.

FAO. 1978. Report on the agro-ecological zones project —Volume 1: Methodology and results for Africa. World Soil Resource Report 48, Food and Agriculture Organization of the United Nations, Rome, Italy. 158 pages and tables.

Hagen, R. M. et al. (Editors). 1967. Irrigation of agricultural lands. Monograph 11, American Society of Agronomy, Madison WI. 1180 pages.

Klingebiel, A. A. and P. H. Montgomery. 1973. Land capability classification. Agriculture Handbook 210, Soil Conservation Service, U.S. Dept. of Agriculture, U.S. Government Printing Office, Washington, DC. 21 pages.

Nelson, L. A. et al. 1963. Detailed land classification— Island of Oahu. Land Study Bureau Bulletin 3, University of Hawaii, Honolulu. 141 pages.

Obeng, H. B. 1968. Land capability classification of the soils of Ghana under practices of mechanized and hand cultivation for crop and livestock production. Transactions of the 9th International Congress of Soil Science 4:215–223.

Olson, G. W. 1974. Land classifications. Volume 4, Number 7. Search (Agriculture, Agronomy 4). Agricultural Experiment Station, New York State College of Agriculture and Life Sciences, Cornell University, Ithaca, NY. 34 pages.

Olson, K. R. and G. W. Olson. 1981. Utilization of soils information in the preferential tax assessment of agricultural land in the United States. Cornell Agronomy Mimeo 81–39, Cornell University, Ithaca, NY. 23 pages.

Pierce, F. J., W. E. Larson, R. H. Dowdy, and W. Graham. 1982. Productivity of soils: Assessing long-term changes due to erosion. Paper presented and reproduced for 37th Annual Meeting of Soil Conservation Society of America, New Orleans, LA, 8–11 Aug. 1982. 38 pages. (Published in Journal of Soil and Water Conservation 38:39–44, 1983.)

Richardson, M. L., S. D. Clemmons, and J. C. Walker. 1979. Soil survey of Santa Cruz and parts of Cochise and Pima Counties, Arizona. Soil Conservation Service and Forest Service, U.S. Dept. of Agriculture and Arizona Agricultural Experiment Station, U.S. Government Printing Office, Washington, DC. 105 pages and 122 soil map sheets.

SCS. 1969. Know your land: Narrative guide, with photographs and captions, and color slide set. Soil Conservation Service, U.S. Dept. of Agriculture, U.S. Government Printing Office, Washington, DC. 12 pages.

Small, E. (Editor) et al. 1982 (reprinted). Climate and soil requirements for economically important crops in Canada. Research Branch, Agriculture Canada, Ottawa. 55 pages.

Storie, R. E. 1964. Handbook of soil evaluation. University of California, Berkeley, CA. 225 pages.

Weiers, C. J. and I. G. Reid. 1974. Soil classification, land valuation and taxation: The German experience. Miscellaneous Study 1, Centre for European Agricultural Studies, Wye College (University of London), Ashford, Kent, England. 37 pages.

Wohletz, L. R. and E. F. Dolder. 1952. Know California's land: A land capability guide for soil and water conservation in California. Dept. of Natural Resources, State of California, Sacramento and Soil Conservation Service, U.S. Dept. of Agriculture, Davis, CA. 43 pages and map.

FIGURE 46/*Landscape showing capability classes in California (adapted from Wohletz and Dolder, 1952).*

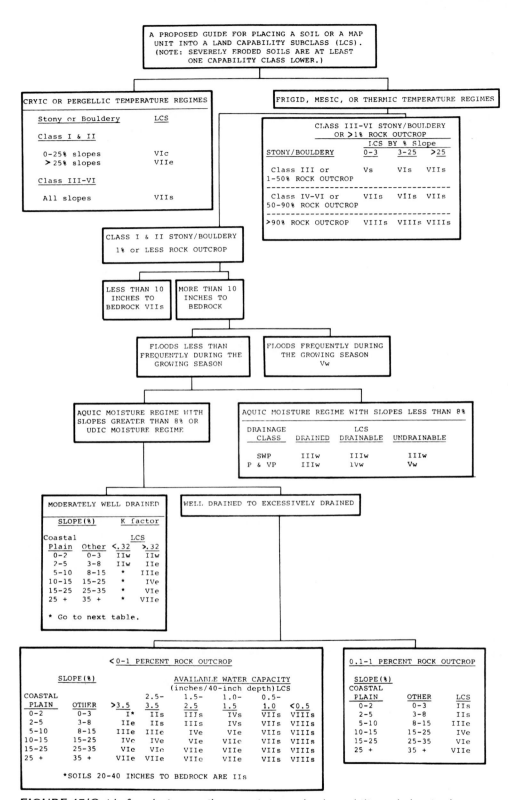

FIGURE 47/*Guide for placing a soil map unit into a land capability subclass in the northeastern United States.*

FIGURE 48/*Irrigated soils in the valley of the Platte River near North Platte, Nebraska.*

FIGURE 49/*Wet soils in need of drainage near New Orleans, Louisiana.*

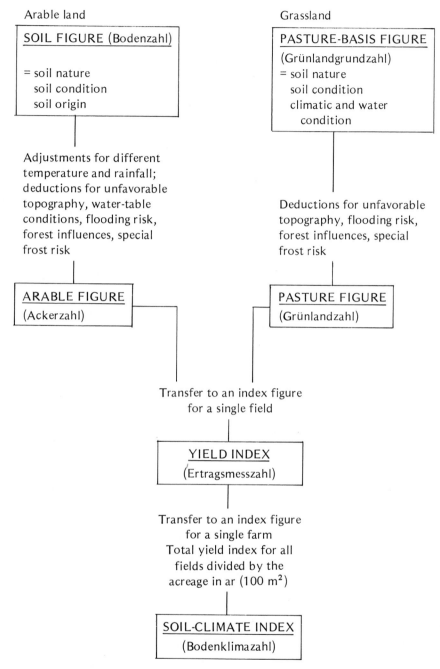

Arable land

SOIL FIGURE (Bodenzahl)

= soil nature
  soil condition
  soil origin

Grassland

PASTURE-BASIS FIGURE
(Grünlandgrundzahl)
= soil nature
  soil condition
  climatic and water
    condition

Adjustments for different
temperature and rainfall;
deductions for unfavorable
topography, water-table
conditions, flooding risk,
forest influences, special
frost risk

Deductions for unfavorable
topography, flooding risk,
forest influences, special
frost risk

ARABLE FIGURE
(Ackerzahl)

PASTURE FIGURE
(Grünlandzahl)

Transfer to an index figure
for a single field

YIELD INDEX
(Ertragsmesszahl)

Transfer to an index figure
for a single farm
Total yield index for all
fields divided by the
acreage in ar (100 m²)

SOIL-CLIMATE INDEX
(Bodenklimazahl)

FIGURE 50/*Outline of the calculation of the soil-climate index (adapted from Weiers and Reid, 1974).*

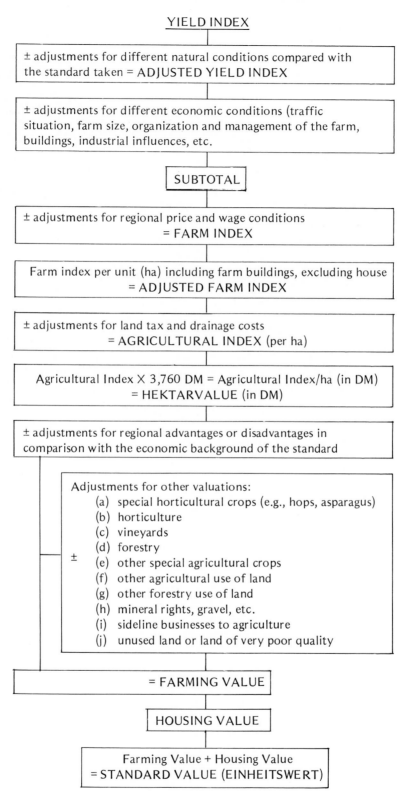

YIELD INDEX

± adjustments for different natural conditions compared with the standard taken = ADJUSTED YIELD INDEX

± adjustments for different economic conditions (traffic situation, farm size, organization and management of the farm, buildings, industrial influences, etc.)

SUBTOTAL

± adjustments for regional price and wage conditions = FARM INDEX

Farm index per unit (ha) including farm buildings, excluding house = ADJUSTED FARM INDEX

± adjustments for land tax and drainage costs = AGRICULTURAL INDEX (per ha)

Agricultural Index X 3,760 DM = Agricultural Index/ha (in DM) = HEKTARVALUE (in DM)

± adjustments for regional advantages or disadvantages in comparison with the economic background of the standard

± Adjustments for other valuations:
    (a) special horticultural crops (e.g., hops, asparagus)
    (b) horticulture
    (c) vineyards
    (d) forestry
    (e) other special agricultural crops
    (f) other agricultural use of land
    (g) other forestry use of land
    (h) mineral rights, gravel, etc.
    (i) sideline businesses to agriculture
    (j) unused land or land of very poor quality

= FARMING VALUE

HOUSING VALUE

Farming Value + Housing Value = STANDARD VALUE (EINHEITSWERT)

FIGURE 51/*Outline of the scheme of the standard valuation process for adjustments beyond the yield index stage (adapted from Weiers and Reid, 1974).*

FIGURE 52/*Woodland harvesting in Fulton County, New York State, at the southern edge of the Adirondack Mountains—on soils that are marginal for farming.*

TABLE 43/*Soil properties (data) important in some different land classification systems (adapted from Beatty et al., 1979).*

| Data | System Author | | | | | |
	Storie (1964)	Klingebiel and Montgomery (1973)	Canada (Anon., 1972a)	Beek and Bennema (1972a)	Obeng (1968)	Bartelli (1968)
Texture	X	X			X	
Drainage	X	X	X	X	X	X
Alkali	X	X	X	X		
Salts	X	X	X	X		
Nutrients	X		X	X	X	X
Acidity	X		X			
Erosion	X	X	X	X	X	
Soil depth	X	X	X		X	X
Permeability		X	X		X	
Available water		X	X	X	X	X
Flooding		X	X			
Toxic substances		X	X			
Stoniness		X	X			

TABLE 44/Criteria for placing soil map units into land capability classes in California (adapted from Beatty et al., 1979).

Capability Class	Effective Soil Depth (in.)	Surface Layer Texture Irrigated	Surface Layer Texture Dryland	Permeability	Drainage Class	Available Water Capacity	Slope (%) Irrigated	Slope (%) Dry	Erosion Hazard	Flooding Hazard	Salinity EC × 10³ at 25°C (mmhos)	Alkali	Toxic Substances	Frost-free Season (days)
I	>40	Moderately coarse, medium moderate slow	Moderately coarse, medium moderate slow	Moderately slow to moderately rapid	Well or moderately well >60 in.	>7.5 in. Average AWC >0.13 in./in. Surface foot AWC >0.13 in./in.	0–2	0–5	None or slight	None or rare	<4 (none)	None	None	>140
II	>40	Coarse (loamy sand) or loamy fine sand to fine (<60% clay) may be gravelly	Moderately coarse, medium, moderately fine and fine (<60% clay)	Rapid through slow	Somewhat poorly through somewhat excessive >36 in.	>5 inch Average AWC >0.08 in./in.	0–5	0–9	None through moderate	None through occasional	<8 (none or slight)	None to slight	None to slight	>100
III	>20	Any, may be gravelly or cobbly	Moderately coarse, medium, moderately fine & fine may be gravelly or cobbly	Rapid through very slow	Poorly through excessive >20 in.	>3.75 in. Average AWC >0.06 in./in.	0–9	0–15	None through high	None through occasional	<16	None to moderate	None to moderate	>80
IV	>10	Any, may be very gravelly or cobbly or stony.	Coarse (loamy sand) or loamy fine sand to fine may be gravelly, cobbly or stony	Any	Poorly through excessive >20 in.	>2.5 in. Average AWC >0.04 in./in.	0–15	0–30	None through very high	None through occasional	<16	None to moderate	None to strong	>50
V	>20	Any and very stony or very cobbly	Any and very stony or very cobbly	Any	Somewhat excessive through very poorly	>2.5 in. Average AWC >0.06 in./in. Irrigated: Root zone >5.0 in. average AWC >0.08 in./in. Nonirrigated	0–15	0–15	None to slight	None through frequent	<8	None or slight	None or strong	>80
VI	>10	Any and very stony or very cobbly	Any and very stony or very cobbly	Any	Any	>2.5 in. Average AWC >0.06 in./in.	0–30	0–50	None to high	None through frequent	All— Irrigated <8 none or slight dryland	All Irrigated or slight dryland	None to strong	>50
VII	Any	Any	Any	Any	Any	>1 in.	Any	<75	None to very high	Any	Any	Any	Any	Any
VIII	Any	Any	Any	Any	Any	Any	Any	Any	Any	Any	Any	Any	Any	Any

TABLE 45/*Percentage values assigned to factors* A, B, C, X, *and* Y *in determining land productivity indexes on Oahu (adapted from Nelson et al., 1963).*

Factor A — general character of the soil profile:

Immature soils with little profile development (alluvial soils and soils formed under the influence of excess moisture in the profile).

Deep, well-drained soils	92–100%
Deep, moderately well drained soils	80– 91%
Moderately deep, well-drained soils	90– 95%
Moderately deep, moderately well drained soils	71– 85%
Moderately deep, imperfectly to poorly drained soils	60– 70%
Shallow, moderately well drained soils	55– 65%
Deep, imperfectly to poorly drained soils	20– 55%

Moderately well developed to well-developed upland soils from basalts, andesites, volcanic ash or cinders, or alluvium.

Deep, well-drained soils	92–100%
Deep, moderately well drained soils	85– 94%
Deep, imperfectly to poorly drained soils	75– 84%
Moderately deep, well-drained soils	90– 95%
Moderately deep, moderately well drained soils	71– 85%
Moderately deep, imperfectly to poorly drained soils	60– 70%
Shallow, well-drained soils	70– 80%
Shallow, moderately well drained soils	60– 69%
Shallow, well-drained, eroded soils	40– 50%
Shallow, imperfectly to poorly drained soils	30– 39%
Water-logged lands	25– 65%
Lands in which bedrock is exposed at the surface	10– 24%

Lithosols and Regosols

Shallow, well-drained soils developed from aa lava or volcanic cinders and ash (soils may be cultivated, but only with difficulty)	70– 85%
Deep, excessively drained coral or basaltic sands	25– 45%
Lands in which stones and rocks are exposed at the surface (practically no soil)	0– 40%

Man-made lands

Moderately deep to deep, well-drained fill materials	80– 95%
Shallow, well-drained fill materials	60– 70%

Factor B — Texture of surface soil:

Silt loam and fine sandy loam	90–100%
Silty clay loam, clay loam, and silty clay (mainly nonexpanding 1:1 type minerals and metallic oxides)	85– 98%
Plastic clay (mainly expanding 2:1 type minerals; characteristic granular structure)	82– 92%
Loamy sand, sandy loams	85– 95%
Coarse sand, medium sand	65– 75%

Miscellaneous situations

Stony lands (including aa)	65– 85%
Rocky lands	25– 50%
Pahoehoe	20– 40%

Factor C — Slope of land:

0–10%	100%
11–20%	90%
21–35%	75%
36–80%	50%
>80%	15%

Factor X — Miscellaneous factors:

Reaction of surface soil

Medium acid to mildly alkaline (pH 5.6–7.5)	90–100%
Alkaline (pH>7.5)	85– 89%
Acid (pH<5.5)	80– 89%

Salinity

None to slight (little or no interference of soluble salts with normal crop growth)	86–100%
Moderate (soluble salts interfere considerably with normal crop growth)	75– 85%
Severe (soils have excess of soluble salts, mainly NaCl, which prevent the normal growth of ordinary crop plants)	55– 65%

Fertility level (as measured by modified Truog quick soil tests)

Ample (>125 lb. P_2O_5/A; >240 lb. K_2O/A; >4000 lb. CaO/A)	95–100%
Moderate (50–125 lb. P_2O_5/A; 80–240 lb. K_2O/A; 1000–4000 lb. CaO/A)	85– 94%
Poor to very poor (<50 lb. P_2O_5/A; <80 lb. K_2O/A; <1000 lb. CaO/A)	65– 84%

Erosion

None to slight (<25% of original surface soil removed from most of area)	95–100%
Moderate (25–50% of original surface soil removed from most of area)	90– 94%
Severe (practically all of original surface soil removed from most of area)	85– 89%

Winds

Slight (maximum velocities <31 mph)	95–100%
Moderate (maximum velocities 32–50 mph)	90– 94%
Severe (maximum velocities >50 mph)	85–89%

Factor Y — Average annual rainfall:

0–20 inches	55– 79%
21–40 inches	90– 94%
41–60 inches	85– 98%
61–90 inches	80– 84%
91–150 inches	75– 79%
>150 inches	70– 74%

TABLE 46/*General soil characteristics for Class 1, 2, and 3 of irrigation suitability (adapted from Beatty et al., 1979).*

Soil Characteristics	Class 1 — Arable	Class 2 — Arable	Class 3 — Arable
Soil			
Texture	Sandy loam to friable clay loam	Loamy sand to very permeable clay	Loamy sand to permeable clay
Depth			
To sand, gravel, or cobble	90 cm plus of fsl or finer or 105 cm of sl	60 cm plus fsl or finer 75 cm sl to ls	45 cm plus fsl or finer or 60 cm of coarser soil
To shale or similar material 15 cm less to rock	150 cm plus 135 cm with minimum of 15 cm gravel overlying impervious material or sandy loam throughout	120 cm plus or 105 cm with minimum of 15 cm gravel	105 cm plus or 90 cm with minimum of 15 cm of gravel overlying impervious material or loamy sand throughout
To penetrable lime horizon	45 cm with 150 cm penetrable pH is < 9.0	35 cm with 120 cm penetrable	25 cm with 90 cm penetrable
Alkalinity	Unless soil is calcareous, total salts are low and evidence of black alkali is absent	pH 9.0 or less unless soil is calcareous, salts are low, and evidence of black alkali is absent	pH 9.0 or less unless soil is calcareous, total salts are low, and evidence of black alkali is absent
Salinity	Total salts not to exceed 0.2%	Total salts not to exceed 0.5%	Total salts not to exceed 0.5%
Topography			
Slopes	Smooth slopes up to 4% in reasonably large size areas	Smooth slopes up to 8% in general gradient	Smooth slopes up to 12% in general gradient
Surface	Even enough to require only small amount of leveling no heavy grading	Moderate grading required but feasible at reasonable cost	Heavy and expensive grading in spots but in amount found feasible
Cover (loose rocks)	Clearing cost small	Sufficient to reduce productivity and interfere with cultural practices, clearing required but at moderate cost	Present in amounts to require expensive but feasible clearing
Drainage			
Soil and topography	Soil and topographic conditions such that no specific farm drainage requirement is anticipated	Some farm drainage will probably be required; reclamation by artificial means feasible at reasonable cost	Significant farm drainage required; reclamation by artificial means expensive but feasible

TABLE 47/*Examples of standard mapping symbols and components for land classification surveys of the U.S. Bureau of Reclamation (adapted from Olson, 1974).*

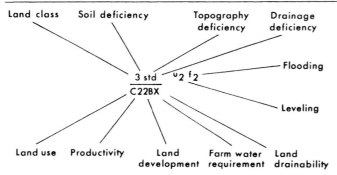

Basic land classes and subclasses:
  Arable Class 1: 1
  Arable Class 2: 2s, 2t, 2d, 2st, 2sd, 2td, 2std
  Arable Class 3: 3s, 3t, 3d, 3st, 3sd, 3td, 3std
  Limited Arable Class 4: pasture—4Ps, 4Pt, 4Pd, 4Pst,
    4Psd, 4Ptd, 4Pstd
      Similar subclasses for fruit 4F, rice 4R, truck 4V,
        suburban 4H, sprinkler 4S, and subirrigation 4U
  Tentatively Nonarable Class 5:
    Pending investigation: 5s, 5t, 5d, 5st, 5sd, 5td, 5std
    Pending reclamation: 5(1), 5(2s), 5(2t), etc.
      Project drainage: 5d(1), 5d(2s), 5d(2t), etc.
      Similar subclasses for flooding: 5f
    Pending investigation or reclamation
      Isolated: 5i(1), 5i(2s), 5i(2t), etc.
      Similar subclasses for high 5h and low 5l
Nonarable Class 6: 6s, 6t, 6d, 6st, 6sd, 6td, 6std
  Isolated: 6i(1), 6i(2s), 6i(2t), etc.
  Similar subclasses for high 6h, low 6l, and water right 6W
  (6W denotes water rights encountered in the classification)
Subclass designations:
  s—soils
  t—topography
  d—farm drainage
Soils appraisals:
  k—shallow depth to coarse sand, gravel, or cobbles
  b—shallow depth to relatively impervious substrata
  z—shallow depth to concentrated zone of lime
  v—very coarse texture (sands, loamy sands)

l—moderately coarse texture (sandy loams, loams)
m—moderately fine texture (silt loams, clay loams)
h—very fine texture (clays)
e—structure
n—consistence
q—available moisture capacity
i—infiltration
p—hydraulic conductivity
r—stoniness
y—soil fertility
a—salinity and alkalinity
Topographic appraisals:
  g—slope
  u—surface
  j—irrigation pattern
  c—brush or tree cover
  r—rock cover
Drainage appraisals:
  f—surface drainage—flooding
  w—subsurface drainage—water table
  o—drainage outlet
Land use:
  C—irrigated cultivated
  L—nonirrigated cultivated
  P—irrigated permanent grassland
  G—nonirrigated permanent grassland
  B—brush or timber
  H—suburban or homestead
  W—waste or miscellaneous
ROW—right of way
  Productivity and land development:
    1, 2, 3, 4, or 6 denote land class level of factor, such as:
      Class 2 productivity, Class 2 development cost—"22"
  Farm water requirement:
    A—low
    B—medium
    C—high
  Land drainability:
    X—good
    Y—restricted
    Z—poor or negligible

TABLE 48/*Summary of the major features of agricultural land tax assessment methods used in the United States (adapted from Olson and Olson, 1981).*

Feature	Alabama	Alaska	Arizona	Arkansas	California	Colorado	Connecticut	Delaware	Florida	Georgia	Hawaii	Idaho	Illinois	Indiana	Iowa	Kansas	Kentucky	Louisiana	Maine	Maryland	Massachusetts	Michigan	Minnesota	Mississippi	Missouri	Montana	Nebraska	Nevada	New Hampshire	New Jersey	New Mexico	New York	North Carolina	North Dakota	Ohio	Oklahoma	Oregon	Pennsylvania	Rhode Island	South Carolina	South Dakota	Tennessee	Texas	Utah	Vermont	Virginia	Washington	West Virginia	Wisconsin	Wyoming	Total
Uses soil survey maps	X	X	X	X	X	X				X		X	X	X	X	X	X	X	X	X	X	X	X	X	X	X	X			X		X		X				X			X	X	X	X	X	X	X	X		X	34
Land capability classes used				X		X										X								X		X																			X	X	X				6
Soil productivity classes used									X	X	X		X	X	X		X					X	X							X		X				X	X							X					X		11
Combination of productivity and capability	X	X		X	X							X	X	X	X			X		X					X		X			X				X			X	X			X		X	X			X			X	18
Soil properties used to rate soils				X																		X													X	X	X	X											X		6
Statewide productivity ranking of soils																									X		X			X	X																				8
Any computerized soil maps																						X	X											X	X																5
Soil survey report yields	X	X		X		X			X	X		X	X	X	X		X	X	X	X	X	X	X	X	X	X	X			X		X			X			X		X	X		X	X						X	27
Experiment station research yields				X								X	X	X	X							X	X	X	X	X	X			X	X	X			X	X	X			X				X							11
Crop reporting service yield		X		X								X	X	X	X							X	X			X	X			X	X	X			X	X	X			X											14
Statewide yield estimates												X	X	X	X								X			X	X							X	X	X				X	X										10
Yields based on collective judgments		X										X	X	X	X							X	X		X		X					X		X	X		X														12
Yield estimates for more than one crop	X												X									X										X		X			X														5
Fair market value			X				X	X															X						X	X			X					X				X									9
Preferential treatment for farmland	X	X	X	X	X	X		X	X	X	X	X	X	X	X	X	X	X	X	X	X	X	X	X	X	X	X	X	X	X	X	X	X	X	X	X	X	X	X	X	X		X	X	X	X	X	X	X	X	41
Use value	X	X	X	X	X			X	X	X	X	X	X	X	X	X	X	X	X	X	X	X	X	X	X	X	X	X	X	X	X	X	X	X	X	X	X	X	X	X	X		X	X	X	X	X	X	X	X	41
Reduced assessment ratios																																				X				X											2
Circuit breaker provisions				X								X		X																											X								X		4
Property tax deferral	X	X	X	X	X				X	X						X	X	X	X		X			X	X	X	X			X		X			X	X	X	X	X		X		X	X	X	X	X	X	X	X	24
Contractual agreements	X	X	X		X				X	X						X	X	X	X		X	X				X	X		X				X	X			X	X	X	X		X	X	X	X	X	X	X	X		19
Income capitalization approach	X	X	X	X	X				X	X				X	X		X	X	X		X	X	X	X		X			X	X	X	X	X	X	X	X	X	X	X	X	X	X	X	X	X	X	X	X	X	X	34
Sales or market data approach		X		X																						X	X		X		X		X	X			X	X		X				X	X	X	X	X	X	16
Comparisons income vs. sales																						X									X		X	X				X									X	X			8
Comparison productivity vs. sales	X										X	X	X	X								X	X								X		X	X	X															X	10

TABLE 49/Computer list of some of the soil map units in New York State with lime level, rotation, crop and TDN yields, and mineral soil group for tax assessment.

Observation	Map Symbol	Slope	Soil	Texture	Modifier	County	Lime (1 = A 2 = B)	Rotation (yrs corn in 10 yrs)	Silage Corn (T/A)	Hay (T/A)	TDN Yields (T/A)	Mineral Soil Group
2366	Ha	03	HALSEY	L		WYOM	2	2	15	2.0	1.40	7
2367	199A	03	HALSEY	MKL		FULT	2	2	15	2.0	1.40	7
2368	96	03	HALSEY	MKL		JEFF	2	2	15	2.0	1.00	8
2369	Ha	03	HALSEY	MKL		ONON	2	2	15	2.0	1.40	7
2370	39	03	HALSEY	MKSIL		CATT	2	2	15	2.0	1.40	7
2371	39	03	HALSEY	MKSIL		CHAU	2	2	15	2.0	1.40	7
2372	26A	03	HALSEY	MKSIL		COLU	2	2	15	2.0	1.40	7
2373	213A	03	HALSEY	MKSIL		DUTC	2	2	15	2.0	1.40	7
2374	Ha	03	HALSEY	MKSIL		SCHO	2	2	15	2.0	1.40	7
2375	Hc	03	HALSEY	MKSIL		TOMP	2	2	15	2.0	1.40	7
2376	Ha	03	HALSEY	MKSIL		WASH	2	2	15	2.0	1.40	7
2377	27A	03	HALSEY	SIL		DELA	2	2	15	2.0	1.40	7
2378	Ha	03	HALSEY	SIL		ERIE	2	2	15	2.0	1.40	7
2379	Wd	03	HALSEY	SIL		LIVI	2	2	15	2.0	1.40	7
2380	Ha	03	HALSEY	SIL		MADI	2	2	15	2.0	1.40	7
2381	Ha	03	HALSEY	SIL		ORAN	2	2	15	2.0	1.40	7
2382	97	03	HALSEY	SIL		OTSE	2	2	15	2.0	1.40	7
2383	153	03	HALSEY	SIL		SARA	2	2	15	2.0	1.40	7
2384	Ha	03	HALSEY	SIL		TOMP	2	2	15	2.0	1.40	7
2385	Ha	03	HALSEY	SIL		WAYN	2	2	15	2.0	1.40	7
2386	HaA	04	HALSEY	SIL		GENE	2	2	15	2.0	1.40	7
2387	CbA	02	HAMLIN	SIL		LEWI	2	7	26	6.0	4.54	1
2388	GdA	02	HAMLIN	SIL		LEWI	2	7	26	6.0	4.54	1
2389	He	03	HAMLIN	FSL		HERK	2	7	26	6.0	4.54	1
2390	200	03	HAMLIN	SIL		ALBA	2	7	26	6.0	4.54	1
2391	2	03	HAMLIN	SIL		CATT	2	7	26	6.0	4.54	1
2392	2	03	HAMLIN	SIL		CHAU	2	7	26	6.0	4.54	1
2393	10	03	HAMLIN	SIL	HI BOTTOM	CHEN	2	7	26	6.0	4.54	1
2394	5	03	HAMLIN	SIL		CHEN	2	7	26	6.0	4.54	1
2395	Hm	03	HAMLIN	SIL		ERIE	2	7	26	6.0	4.54	1
2396	Hf	03	HAMLIN	SIL		HERK	2	7	26	6.0	4.54	1
2397	2	03	HAMLIN	SIL		JEFF	2	7	26	6.0	4.54	1
2398	Hb	03	HAMLIN	SIL		MADI	2	7	26	6.0	4.54	1
2399	Ge	03	HAMLIN	SIL		MONR	2	7	26	6.0	4.54	1
2400	Hc	03	HAMLIN	SIL		MONR	2	7	26	6.0	4.54	1
2401	Ha	03	HAMLIN	SIL		MONT	2	7	26	6.0	4.54	1
2402	Ha	03	HAMLIN	SIL		NIAG	2	7	26	6.0	4.54	1
2403	2	03	HAMLIN	SIL		ONEI	2	7	26	6.0	4.54	1
2404	Hb	03	HAMLIN	SIL		ONON	2	7	26	6.0	4.54	1
2405	Hc	03	HAMLIN	SIL	HI BOTTOM	ONON	2	7	26	6.0	4.54	1
2406	Ha	03	HAMLIN	SIL		ORLE	2	7	26	6.0	4.54	1
2407	1	03	HAMLIN	SIL		OTSE	2	7	26	6.0	4.54	1
2408	HaA	03	HAMLIN	SIL		RENN	2	7	26	6.0	4.54	1
2409	Ha	03	HAMLIN	SIL		SCHE	2	7	26	6.0	4.54	1
2410	Ha	03	HAMLIN	SIL		ULST	2	7	26	6.0	4.54	1
2411	Hb	03	HAMLIN	SIL		WASH	2	7	26	6.0	4.54	1
2412	Hm	03	HAMLIN	SIL		WAYN	2	7	26	6.0	4.54	1
2413	Hc	03	HAMLIN	SIL		WYOM	2	7	26	6.0	4.54	1
2414	Gb	05	HAMLIN	SIL		LIVI	2	7	26	6.0	4.54	1
2415	Rd	05	HAMLIN	SIL	HI BOTTOM	LIVI	2	7	26	6.0	4.54	1
2416	Ga	10	HAMLIN	FSL		LIVI	2	7	26	6.0		1
2417	050AB	08	HAPLAQUEPTS		X BOULDERY	GREE	.	.	.	.		9
2418	HcA	02	HARTLAND	VFSL		WASH	2	7	26	5.5	4.46	1
2419	88A	03	HARTLAND	SIL		CHEN	2	7	26	5.5	4.46	1
2420	49A	03	HARTLAND	VFSL		WARR	2	7	26	5.5	4.46	1

TABLE 50/*Yields of total digestible nutrients (TDN) in index form for mineral soil groups (TDN) factor 0.2 × tons/acre weight for corn silage, 0.5 × tons/acre weight for hay) plus woodland soil classes plus organic soil groups.*

Mineral Soil Group (Cropland)	Index (Productivity Rank)
1	90–100
2	80–89
3	70–79
4	60–69
5	50–59
6	40–49
7	25–39
8	< 24
9	
10	Other
	Rock, marsh, bog, pits, etc.

Woodland Class	Yield
1	$> 10{,}000$ BF/acre
2	2000–10,000 BF/acre
3	< 2000 BF/acre

Organic Soil Group	Definition
A	> 51 in. drained
B	16–51 in. drained
C	> 51 in. wet
D	Shallow wet

TABLE 51/*Soil value (average value per acre) and yields (tons per acre) of corn and hay for mineral soil groups.*

Mineral Soil Group (Cropland)	Value (Average)	Corn (tons/acre)	Hay (tons/acre)
1	$795	18.3	3.8
2	650	17.0	3.5
3	480	15.1	3.1
4	260	12.9	2.6
5	145	11.9	2.3
6	55	10.3	1.9
7	50	8.8	1.5
8	50	Hay	1.0

TABLE 52/*Generalized yields in relation to the soil-climate index (adapted from Weiers and Reid, 1974)*

Soil-Climate Index (points)	Depth of Roots (cm)	Wheat (tons/ha)	Rye (tons/ha)	Sugar Beet (tons/ha)	Potatoes (tons/ha)	Maize (tons/ha)
90	90	9	—	60	40	10
	60	6	—	70	40	10
70	70	7	6	50	30	8
	45	4.5	5	50	30	8
50	50	5	5	40	30	6
	30–50	—	—	40	—	5
30	40	—	4.5	—	30	6
	25	—	3.5	—	25	5

TABLE 53/*Sample of farms for sale during the DLG Agricultural Show, September 1974 (adapted from Weiers and Reid, 1974).*

Hildesheim area:	95 ha, sugar beet, wheatland, good buildings, 75 BP[a] (yield index), 2.5m[b] DM
Einbeck area:	48.75 ha, without buildings, 65 BP, 1.25m DM
Near Paderborn:	100 ha (15 ha forest), 50 BP, 2.00m DM
Palatinate (Pfalz):	110 ha (75 ha arable, 20 ha pasture, 15 ha forest), good buildings, 50–55 BP, 2.50m DM
Wolfenbüttel area:	75 ha, all first-class arable, good buildings, top-class sugar-beet farm, 81 BP, 3.20m DM
Uelzen area:	70 ha (50 arable, rest grass + forest), 50 BP, 0.95m DM
Gifhorn area:	85 ha (55 ha arable, rest pasture), good buildings, 45 BP, 1.70m DM
Hildesheim:	85 ha, all arable, sugar-beet farm, 85 BP, 3.20m DM

[a] BP (Bodenpunkte) = yield index.
[b] m = 1,000.

Erosion control

PURPOSE Control of soil erosion is probably the single most important and extensive application of soil survey interpretations. Areas inadequately vegetated and protected are prone to soil losses through runoff and erosion, and fertility and productivity losses result (Olson, 1979). Relative erosion effects on wheat yields were illustrated in Idaho in 1980 and 1981 with yields of 45.2 bushels per acre on plots with 6 inches of topsoil added, 26.8 bushels per acre on plots left undisturbed, 14.4 bushels per acre on plots with 6 inches removed, and 10.6 bushels per acre on plots with 12 inches removed (Norris and Comis, 1982). In 1977, more than 2 billion tons of soil were lost through sheet and rill erosion in the United States (SCS, 1980a). In New York State, more than 5 tons of soil per acre per year are eroding on 1 million acres of cropland (SCS, 1980b); half of the original topsoil has been lost from sloping soils in the 200 years that they have been farmed. Of the 1,000 farms that are abandoned each year in New York State, many are abandoned at least in part due to soil deterioration and yield declines due to erosion (Cornell University, 1981). Erosion is even more severe in many countries in Africa, Asia, and Latin America (Eckholm, 1976; El-Swaify et al., 1982; FAO, 1980; Kussow et al., 1982), and the ultimate consequences will be serious because of the remorseless workings of the concept of the "tragedy of the commons" (Hardin, 1968), which relates resource management (or mismanagement) to the well-being (or detriment) of a society and civilizations (Brown, 1982; Clawson et al., 1971). The purpose of this exercise is to quantify the control aspects of the problem, and to give students some experience in calculating and predicting erosion losses. The references cited in this section and pages 105–118 of the textbook should be carefully studied in preparation for activities relating to erosion control.

BACKGROUND Extensive literature compilations on soil erosion and yield declines have recently been made (Beatty, 1982; Williams, 1981). It appears that (in general) each centimeter of soil loss from the surface results in more than 1 percent loss in total productivity of a soil, with the percentage yield reduction increasing as a larger amount of the surface topsoil is eroded away (Higgins and Kassam, 1981). From an excellent discussion of the problems of relating potential productivity to soil loss, Figure 53 illustrates some observed and computed percentage productivity losses as a soil is eroded.

Many soils have about 50 percent loss in productivity from a loss of 25 centimeters of topsoil due largely to losses of organic matter and other nutrients and deterioration of physical soil properties. If one assumes that profile removal (to 140 cm) would result in 100 percent loss in productivity, then plots like those in Figure 54 can be constructed (Higgins and Kassam, 1981). Figure 54 also shows a plot for a soil where 50 percent productivity was presumed associated with a lesser depth of topsoil (as for a highly leached soil), and a plot for a soil where 50 percent productivity was presumed associated with a greater depth of topsoil (as for a soil developed in loess). Such curves can be constructed for any depth of soil assuming a 100 percent loss in productivity as a result of the removal of all soil depth (Higgins and Kassam, 1981).

Figure 55 shows results of plotting soil and productivity losses from several soils of the tropics (Higgins and Kassam, 1981). Loss of the top 5 centimeters of soil from a Chromic Vertisol (FAO classification) results in a relatively low productivity loss (5 percent), a loss of 15 percent in a Humic Ferralsol, and a 40 percent loss in a Plinthic Acrisol (Higgins and Kassam, 1981). The characteristics of each profile, obviously, determine the losses in each landscape situation.

Pierce et al. (1982) evaluated erosion effects on productivity of contrasting soils, as shown in Figure 56. Soils with uniform horizons of high productivity retain relatively high yields (A) even when eroded—but such soils are relatively uncommon. Soils with unfavorable subsoils (B) have increasing yield declines after the surface horizons are eroded. The poorest soils (C) with hardpan, rock, or coarse fragments have a sharp yield decline after the surface topsoils are removed. When eroded, soils are difficult to reclaim; it is far more efficient to reduce erosion to a minimum to preserve productivity than to attempt to restore the topsoil after it is gone.

CORNELL TEACHING Recently, the soil-loss equation has been expanded to be more inclusive in a handbook prepared by Wischmeier and Smith (1978). Troeh et al. (1980) have assembled a comprehensive textbook on soil and water conservation for productivity and environmental protection. These references are among those used in teaching soil conservation at Cornell University. For many years M. G. Cline and G. W. Olson have taught the application of the soil loss equation in the Ithaca area using the following format. Students and teachers should adapt this organizational format or similar ratings of soils to fit their soils and environmental characteristics and management situations (e.g., rotations) in their study areas. These Cornell teaching materials help to outline the soil-loss prediction processes, and show how different rotations and soils result in tolerable or intolerable soil losses.

The soil-loss prediction equation is

$$A = RKLSCP$$

where

A = average *annual soil loss* in tons per acre

R = a *rainfall-erosion index* reflecting the contribution of intensity and duration of rainfall

K = a *soil factor* expressed as tons of soil loss per acre per unit of rainfall-erosion index on a 9 percent slope 72.6 feet long under continuous fallow with up–down slope tillage for a specific kind of soil

C = a *cropping-management* factor expressed as the ratio of loss under specified cropping system to the loss under fallow

L = a *length-of-slope factor*, expressed as the ratio of soil loss from a specified length of slope to that defined for the K factor

S = a *slope-gradient factor* expressed as the ratio of soil loss for a specified % slope to that of the K factor

P = a *water-control-practice* factor expressed as the ratio of soil loss with a specified practice to that of up–down slope operation when other factors are equal

In effect, A = R represents average annual potential for the soil-loss characteristic of the rainfall of a given place. A = RK introduces tons of soil loss per acre *per unit of R*, where K is characteristic of each kind of soil and is arbitrarily defined for a set of standard cropping, water control, slope gradient, and slope length conditions, or where

Cropping is fallow.
Water control is none.
Slope gradient is 9 percent.
Slope length is 72.6 feet.

Each of the other factors, C, L, S, and P, is simply expressed as a ratio of the loss under a given condition to the loss under the standard condition for which K is defined. Thus, when these factors are multiplied by RK, they adjust RK to conditions specified (since they are ratios to losses under standard conditions of K).

The system was developed by empirical correlations of observed soil losses in erosion experiments to the conditions under which these experiments were run.

Rainfall Erosion Index—90 for the Ithaca Area Experiments showed that when factors other than rainfall are held constant, soil losses during a given storm were proportional to the product of the total kinetic energy of the water that hit the ground and the maximum intensity for a 30-minute period. When kinetic energy (E) of the storm was expressed as foot-pounds per acre and intensity (I) was expressed in inches per hour the product (EI) was related to soil loss linearly on fallow plots. EI = kA, where k is a constant. The constant k varied with soil type, slope, and other factors, which then had to be treated as separate factors.

It was found also that the annual total soil loss was directly proportional to the sum of EI value of all storms for the year. Thus, the factor R is the mean annual total of EI values of individual storms. These have been calculated for about 2000 stations in the 37 eastern states, and when the data are plotted on a map, lines can be drawn connecting values of equal R (isoerodents). Ithaca is in a zone where R lies between 75 and 100—*about 90.*

Soil Erodibility Factor Soil loss has been measured over a period of years at a number of locations on identified soils where slope gradient, slope length, cropping, and water control practices were defined. These losses, expressed as mean annual losses for fallow plots on 9 percent slopes 72.6 feet long without structures to control runoff, have been divided by the rainfall-erosion index (R) of the site to give K values. These numbers are tons of soil loss per acre per year per unit of R. New

York soils included are Bath flaggy silt loam, 0.05; Dunkirk soil loam, 0.69; Ontario loam, 0.27; and Honeoye silt loam, 0.28.

These studies have also shown that, other conditions being equal, soil loss is related to soil properties such as stones on the surface, texture, permeability, structure, organic matter, and depth. On the basis of similarities of these properties, the soils of the Northeast have been grouped into classes that should be relatively uniform in the erodibility factor. Values of K have been assigned to these groups using the experimentally determined values as benchmarks within an array according to erodibility under standard conditions.

Group 1, K = 0.10. Includes soils in permeable glacial outwash and glacial till and those well-drained soils having slowly permeable substrata but having channery or very gravelly surface soils: Bath, Chenango, Lordstown, Howard, Tunkhannock, and similar soils.

Group 2, K = 0.17. Includes well-drained soils in uniformly sandy gravel-free material: Adams, Windsor, Colonie, and similar soils.

Group 3, K = 0.28. Includes mainly "graded" loams and silt loams with a fragipan or well-expressed textural B over basal till, well to somewhat poorly drained inclusive of soils such as Bath, Erie, Volusia, Morris, Mardin, Langford, Honeoye, Ontario, and similar soils.

Group 4, K = 0.43. Includes "poorly graded" moderately fine and fine textured moderately well and somewhat poorly drained soils in lake and marine sediments: Hudson, Rhinebeck, Schoharie, Odessa, Vergennes, Panton, and similar soils.

Group 5, K = 0.64. Includes "poorly graded" silty or very fine sandy well and moderately well-drained soils in lake sediments: Amboy, Williamson, Dunkirk, Collamer, and similar soils.

Slope Length Factor and Slope Percent Factor
Data from experiments have shown that soil loss is crudely a function of slope length to the 0.5 power.

$$A = f \sqrt{\text{slope length}}$$

Similar data have indicated that soil loss is related to slope percent as a parabolic function. By empirical curve fitting,

$$S = 0.520 + 0.363s + 0.052s^2$$

where s represents percent slope.

These two factors can conveniently be combined to give roughly SL ratios as follows for the Northeast:

Slope (%)	Slope Length (ft)				
	Standard (72.6)	100	200	300	400
3	0.2	0.5	0.8	1.1	1.2
8	0.7	1.0	1.4	1.7	2.0
Standard (9)	1.0	—	—	—	—
15	2.2	2.5	3.7	4.5	5.2
20	2.9	4.1	5.8	—	—

Cropping Factor This factor has been developed on the basis of the effects of different crops at different periods during their culture in relation to the rainfall-erosion indexes for those periods. The values also take into account the effects of differences in cover related to yield (as a measure of growth) of the crop, which is governed in part by fertilization, liming, and similar treatments. Representative average annual soil-loss ratios (relative to standard fallow plot) for well-fertilized, high-yielding rotations would be as follows:

Continuous corn (without cover crop)	0.250
Continuous corn (with cover crop)	0.210
Corn-corn-oats-hay (without cover crop)	0.132
Corn-oats-hay (without cover crop)	0.092
Corn-oats-hay-hay (without cover crop)	0.070
Corn-oats-hay-hay-hay (without cover crop)	0.057
Oats-wheat-hay-hay-hay	0.066
Oats-hay-hay-hay	0.037
Continuous hay	0.004

Depending on assumptions about various kinds and degrees of supporting management practices, these values would require adjustment up or down, but they are close approximations for good farmers of New York.

Water-Control Practice Factor From erosion experiments on a few soils, the following are approximate _P_ values (ratio of soil loss to that of up–down slope tillage).

Slope (%)	P Values		
	Contouring	Strip Cropping	Terracing
1–12	0.50–0.60	0.25–0.30	0.10–0.12
12–18	0.80	0.40	0.16
18–24	0.90	0.45	0.18

Terracing may be treated as a slope-length factor, each terrace breaking the slope into effective run-off areas.

Allowable Soil Loss Erosion is more serious on some soils than on others. It is more serious if root zones are restricted by bedrock or dense layers or if undesirable material is incorporated in the plowed layer after erosion than in deep uniform soil material. On the basis of a number of factors of this kind, tolerances for soil loss without serious deterioration have been estimated for individual soils (Schmidt et al., 1982). Although these vary somewhat among soils within groups of _K_ factors, the following are close approximations for our purposes:

Soil Group 1	3 tons/acre per year
Soil Group 2	5 tons/acre per year
Soil Group 3	2 tons/acre per year
Soil Group 4	3 tons/acre per year
Soil Group 5	4 tons/acre per year

Applications of the Complete Equation at Ithaca ($A = RKLSCP$) Table 54 gives predictions for a sample of combinations of the factors under conditions common at Ithaca. Predicted losses that exceed the tolerances specified are underlined.

For example, Bath channery silt loam has a _K_ value of 0.05 (see page 110 of the textbook). Under continuous corn (C-C-C) it could be safely farmed without excessive erosion on 15 percent slopes of 200-foot length; slopes of 8 percent would be excessive on 400-foot slopes (Table 54). Dunkirk silt loam (_K_ value 0.69) would have excessive soil erosion on all soil map units under continuous corn (Table 54). Dunkirk soils could be safely farmed without excessive erosion, however, on gentler slopes under oats–hay (O-H-H-H) with one year of oats and three years of hay, under corn–oats–hay (C-O-H-H-H), and under the C-O-H and C-C-O-H rotations with contour cultivation

and strip cropping as indicated in Table 54. Students can interpolate values from the references for their soils if data are not available; _K_ values can be estimated from pages 110–111 of the textbook. Each soil map unit should be managed in accord with its capabilities; if erosion is excessive, productive potential will decline. Regional development in different countries is much dependent on control of soil erosion (Smith, 1981). Government programs to provide incentives to farmers to control erosion are vital to preserve the societal requirements for potential productivity of the soils for the future (King, 1982).

TECHNIQUES In the field, slide rules (Figure 57) and electronic calculators can be used to predict soil loss from each soil map unit under different management conditions. Such predictions are used by Conservationists of the Soil Conservation Service in advising farmers about the best crops to grow in individual fields, and in making soil and water conservation plans for each farm. Similar calculations are used in urban construction areas where large areas of bare soils are exposed, and where vegetation must be established quickly before runoff and erosion are excessive. Often, additional engineering constructions are needed, such as sediment basins, catchment dams, and diversion ditches and terraces.

Students with computer facilities should carry the calculations for erosion control into economic considerations if possible. Figure 58 illustrates one example of the application of economics to erosion control. The SCS computer program COSTS (Cost of Soil Treatment Systems) is extremely useful to show the alternatives for practical erosion control. In the display for eroded Honeoye soil, 8 to 15 percent slopes, only the minimal practices (with negative "costs" in Figure 58) are profitable for the farmer. Thus, $1.64 is returned from soil saved per acre per year under contour conservation tillage with a winter cover crop. A continuous cover crop would save $3.00, but conservation tillage in a rotation would save only $0.21. No-till cultivation with winter cover would save $2.00. Contouring, terracing, winter cover, rotation, strip crop, pasture, and retirement from cultivation would be too costly to the farmer. Thus, the COSTS computer program demonstrates quite clearly that the most effective and costly erosion control practices will not be used by the average farmer (acting under economic constraints), unless society (government) provides an incentive for erosion control. In the future, computers will have an increasing role in land-use planning and erosion control (Kling and Olson, 1975).

Students, teachers, and others should carefully observe processes of soil erosion under way in their project areas and in other places. Photographs should be taken to document the changes, especially immediately after intense rainstorms. Whenever the soil is bare, soil erosion is a hazard. During the fall of 1981, for example, an area of several hectares was cleared for arboreteum development on the campus of Cornell University (Olson, 1972). An intense rainstorm caused the erosion, sedimentation, runoff, and flooding illustrated in Figures 59 to 62. Yet, practically nobody, not even developers of the area, realized the full implications of the soil erosion effects. In terms of nutrient losses and environmental degradation, the soil erosion was very damaging indeed. Students, developers, and others, using the soil-loss equation outlined in this exercise, could have calculated and predicted the tons of soil loss that would be removed from each soil map unit in the intense rainstorm of 1981—and in subsequent years under different actual and hypothetical alternative management conditions.

Concepts of soils by most people are two-dimensional (Williams and Ortiz-Solorio, 1981; Williams, 1980). Yet the true landscape perspective of soils is three-dimensional (see the textbook). Through educational processes, it will be necessary to change the perception of soils of most people if productivity and yields are to be maintained. People must be taught that erosion processes are similar in all parts of the world (see Figures 63 to 65), and that the soil-loss equation with adequate modifications can be adapted to help all farmers and all people in all societies. With knowledge of the importance of topsoils as illustrated in Figures 53 to 58, and knowledge about soil movement and runoff as illustrated in Figures 59 to 65, soil erosion losses can be predicted (Table 54) and educational and conservation programs can be planned and implemented accordingly (Wischmeier and Smith, 1978).

REFERENCES

Beatty, M. T. (Chairman). 1982. Soil erosion: Its agricultural, environmental, and socioeconomic implications. Report 92, Council for Agricultural Science and Technology, Ames, IA. 29 pages.

Brown, L. R. 1982. Research and development for a sustainable society. American Scientist 70:14-17.

Clawson, M., H. H. Landsberg, and L. T. Alexander. 1971. The agricultural potential of the Middle East. American Elsevier Publishing Company, Inc., New York. 315 pages and maps.

Cornell University. 1981. Facts and figures. New York State College of Agriculture and Life Sciences, Cornell University, Ithaca, NY. 38 pages.

Eckholm, E. P. 1976. Losing ground: Environmental stress and world food prospects. W. W. Norton & Company, Inc., New York. 223 pages.

El-Swaify, S. A., E. W. Dangler, and C. L. Armstrong. 1982. Soil erosion by water in the tropics. Research Extension Series 024, Dept. of Agronomy and Soil Science, College of Tropical Agriculture and Human Resources, University of Hawaii, Honolulu. 173 pages.

FAO. 1980. In the land dies... Filmstrip of 259 photos. Food and Agriculture Organization of the United Nations, Rome, Italy.

Hardin, G. 1968. The tragedy of the commons. Science 162:1243-48.

Higgins, G. M. and A. H. Kassam. 1981. Relating potential productivity to soil loss. Pages 21-25, in Technical Newsletter of the Land and Water Development Division, for the Field Officer, Food and Agriculture Organization of the United Nations, Rome, Italy. 38 pages.

King, S. S. 1982. Cuts in soil erosion program worry farmers. The New York Times. Sunday, 14 Feb. Page 23.

Kling, G. F. and G. W. Olson. 1975. Role of computers in land use planning. Information Bulletin 88, New York State College of Agriculture and Life Sciences, Cornell University, Ithaca, NY. 12 pages.

Kussow, W. (Chairman of Editorial Committee) et al. 1982. Soil erosion and conservation in the tropics. Special Publication 43, American Society of Agronomy, Madison, WI. 149 pages.

Norris, S. and D. L. Comis. 1982. Soil erosion reduces wheat yields. Soil and Water Conservation News 2:9-10.

Olson, G. W. 1972. Significance of soils to the Cornell Plantations. The Cornell Plantations 28:1-18.

Olson, G. W. 1979. The "state-of-the-art" in linking soil fertility work to the soil survey. (unabridged report). Soil Conservation Service, U.S. Dept. of Agriculture, Washington, DC/Dept. of Agronomy, Cornell University, Ithaca, NY. Reproduced as Cornell Agronomy Mimeo 79-29. 143 pages.

Pierce, F. J., W. E. Larson, R. H. Dowdy, and W. Graham. 1982. Productivity of soils: Assessing long-term changes due to erosion. Paper presented and reproduced for 37th Annual Meeting of Soil Conservation Society of America, New Orleans, LA, 8-11 Aug. 1982. 38 pages. (Published in Journal of Soil and Water Conservation 38:39-44, 1983.)

Schmidt, B. L. (Chairman of Editorial Committee) et al. 1982. Determinants of soil loss tolerance. Special Publication 45, American Society of Agronomy, Madison, WI. 153 pages.

SCS. 1980a. America's soil and water: Condition and trends. Soil Conservation Service, U.S. Dept. of Agriculture, Washington, DC. 32 pages.

SCS. 1980b. New York's vanishing farmland. Soil Conservation Service, U.S. Dept. of Agriculture, Syracuse, NY. 18 pages.

Smith, N. J. H. 1981. Colonization lessons from a tropical forest. Science 214:755-761.

Troeh, F. R., J. A. Hobbs, and R. L. Donahue. 1980. Soil and water conservation for productivity and environmental protection. Prentice-Hall, Inc., Englewood Cliffs, NJ. 718 pages.

Williams, B. J. 1980. Pictorial representation of soils in the Valley of Mexico: Evidence from the Codex Vergara. Pages 51-62 in Geoscience and man (historical geography of Latin America), Vol. XXI, School of Geoscience, Louisiana State University, Baton Rouge. 163 pages.

Williams, J. R. (Chairman). 1981. Soil erosion effects on soil productivity: A research perspective. Journal of Soil and Water Conservation 36:82-90.

Williams, B. J. and C. A. Ortiz-Solorio. 1981. Middle American folk soil taxonomy. Annals of the Association of American Geographers 71:335-358.

Wischmeier, W. H. and D. D. Smith. 1978. Predicting rainfall erosion losses: A guide to conservation planning. Agriculture Handbook 537, Science and Education Administration, U.S. Dept. of Agriculture, U.S. Government Printing Office, Washington, DC. 58 pages.

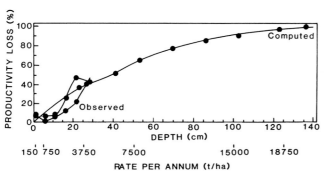

FIGURE 53/*Comparison of computed and observed relationships between soil loss and loss in productivity of Tama silt loam (adapted from Higgins and Kassam, 1981).*

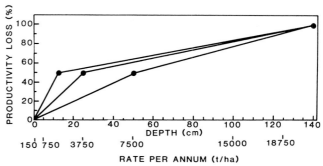

FIGURE 54/*Hypothetical relationships between soil loss and loss in productivity for different soils (adapted from Higgins and Kassam, 1981).*

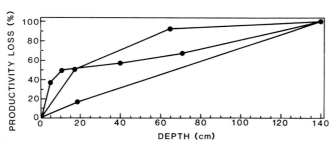

FIGURE 55/*Computed relationships of effects of soil loss on productivity as reflected through organic matter distribution in the profiles for a Chromic Vertisol (Sudan), Humic Ferralsol (Cameroon), and Plinthic Acrisol (Thailand) (adapted from Higgins and Kassam, 1981).*

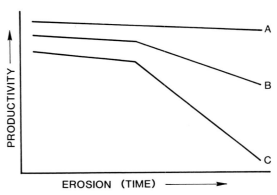

FIGURE 56/*Concept of eroding productivity (adapted from Pierce et al., 1982) for three contrasting soils with: A, favorable surface and subsoil horizons (Monona silt loam); B, favorable surface and unfavorable subsoil horizons (Kenyon loam); C, favorable surface and consolidated or coarse fragment subsoil horizons (Rockton loam).*

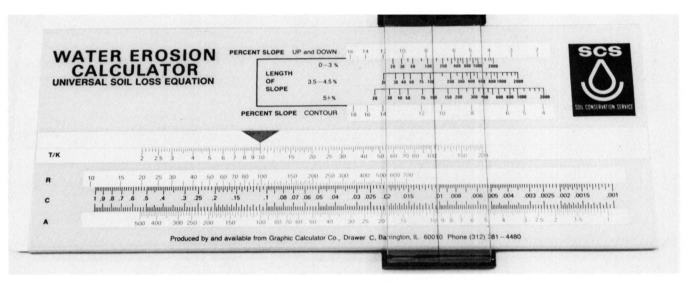

FIGURE 57/*Water erosion calculator in a slide-rule assembly that enables prediction of soil loss from a soil map unit in the field.*

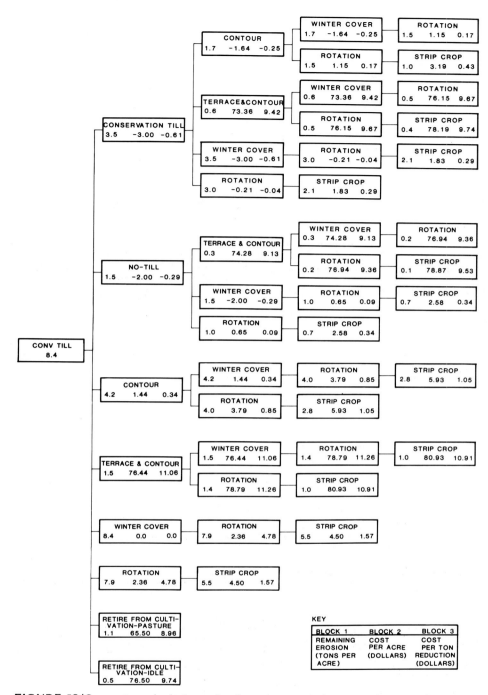

FIGURE 58/*Computer calculations of soil erosion control costs and returns for various management practices on severely eroded Honeoye silt loam, 8 to 15 percent slopes, in Herkimer County, New York. Current expenses and returns of June 2, 1982, are assumed. Each block gives the erosion control practice, remaining erosion after the control practice is installed, cost per acre, and cost per ton reduction of soil loss. The cost values are negative where there is a profit return to the farmer; all other cost values represent costs to the farmer or to society for the various erosion control practices.*

FIGURE 59/*Soil erosion in cornfield at Caldwell Field on the campus of Cornell University, after an intense rainstorm in the fall of 1981.*

FIGURE 60/*Sheet, rill, and incipient gully erosion on a steep slope in the development area of the Cornell Plantations, after an intense rainstorm in the fall of 1981.*

FIGURE 61/*Erosion and deposition in the upper part of the Cornell Plantations development area, after an intense rainstorm in the fall of 1981.*

FIGURE 62/*Flooding in Ithaca, New York, after an intense rainstorm in the fall of 1981.*

FIGURE 63/*Soil disturbance can result in erosion when the surface is bare. Vegetative cover and mulch in this field in the Philippines would help to reduce the soil loss (photo from Rogelio Concepcion).*

FIGURE 64/*Severe erosion on inadequately protected steep slopes in the Philippines (photo from Rogelio Concepcion).*

FIGURE 65/*Contour planting on steep slopes in the Philippines (photo from Rogelio Concepcion).*

TABLE 54/*Predicted average soil loss in tons per acre per year for the indicated combinations of factors,* R = 90.

Cropping System, C	Supporting Practices, P	Slope Length, L					
		200 ft			400 ft		
		3% (.8)	8% (1.4)	15% (3.7)	3% (1.2)	8% (2.0)	15% (5.2)
Soil Group 1; K = 0.10; RK = 9 tons/acre; Allowable Soil Loss 3 tons/acre							
C-C-C (0.21) RKC = 1.9	None (0.0)	1.5	2.6	<u>7.0</u>	2.3	<u>3.8</u>	<u>9.8</u>
	Contour (0.5–0.8)	0.8	1.3	<u>5.6</u>	1.1	1.9	<u>7.8</u>
C-C-O-H (0.13) RKC = 1.2	None (0)	1.0	1.7	<u>4.5</u>	1.4	2.4	<u>6.2</u>
	Contour (0.5–0.8)	0.5	0.8	<u>3.6</u>	0.7	1.2	<u>4.9</u>
	Strips (0.25–0.4)	0.2	0.4	1.8 [a]	0.4	0.6	2.5
C-O-H (0.09) RKC = 0.8	None (0)	0.6	1.1	3.0	1.0	1.6	<u>4.1</u>
	Contour (0.5–0.8)	0.3	0.6	2.4	0.5	0.8	<u>3.3</u>
	Strips (0.25–0.4)	0.2	0.3	1.2 [a]	0.3	0.4	1.6 [a]
C-O-H-H-H (0.06) RKC = 0.5	None (0)	0.4	0.7	1.9 [a]	0.6	1.0	2.6
	Contour (0.5–0.8)	0.2	0.4	1.5 [a]	0.3	0.5	2.1
	Strips (0.25–0.4)	0.1	0.2	0.8 [a]	0.2	0.3	1.0 [a]
O-H-H-H (0.04) RKC = 0.4	None (0)	0.3	0.6	1.5 [a]	0.5	0.8	2.1
	Contour (0.5–0.8)	0.2	0.3	1.2 [a]	0.2	0.4	1.7 [a]
	Strips (0.25–0.4)	0.1	0.2	0.6 [a]	0.1	0.2	0.8 [a]

TABLE 54 (*continued*)

Soil Group 2; $K = 0.17$; $RK = 15.3$ tons/acre; Allowable Soil Loss 5 tons/acre

C-C-C	None	2.3	4.5	<u>11.8</u>	3.8	<u>6.4</u>	16.6
$RKC = 3.2$	Contour	1.1	2.2	<u>9.4</u>	1.9	3.2	<u>8.3</u>
C-C-O-H	None	1.6	2.8	<u>7.4</u>	2.4	4.0	10.4
$RKC = 2.0$	Contour	0.8	1.4	<u>5.9</u>	1.2	2.0	<u>8.3</u>
	Strips	0.4	0.7	3.0[a]	0.6	1.0	4.2
C-O-H	None	1.1	2.0	<u>5.2</u>	1.7	2.8	<u>7.3</u>
$RKC = 1.4$	Contour	0.6	1.0	4.1	0.8	1.4	<u>5.8</u>
	Strips	0.3	0.5	2.1[a]	0.4	0.7	2.9[a]
C-O-H-H-H	None	0.7	1.3	3.3	1.1	1.8	4.7
$RKC = 0.9$	Contour	0.4	0.6	2.6[a]	0.5	0.9	3.8
	Strips	0.2	0.3	1.3[a]	0.3	0.5	1.9[a]
O-H-H-H	None	0.5	0.8	2.2[a]	0.7	1.2	3.1[a]
$RKC = 0.6$	Contour	0.2	0.4	1.8[a]	0.4	0.6	2.5[a]
	Strips	0.1	0.2	0.9[a]	0.2	0.3	1.2[a]

Soil Group 3; $K = 0.28$; $RK = 25$, Allowable Soil Loss 2 tons/acre

C-C-C	None	<u>4.2</u>	<u>7.4</u>	<u>19.6</u>	<u>6.4</u>	<u>10.6</u>	<u>27.7</u>
$RKC = 5.3$	Contour	<u>2.1</u>	<u>3.7</u>	<u>15.7</u>	<u>3.2</u>	<u>5.3</u>	<u>22.1</u>
C-C-O-H	None	<u>2.6</u>	<u>4.6</u>	<u>12.2</u>	<u>4.0</u>	<u>6.6</u>	<u>17.2</u>
$RKC = 3.3$	Contour	1.3	<u>2.3</u>	<u>9.8</u>	2.0	<u>3.3</u>	<u>13.8</u>
	Strip	0.7	1.2	<u>4.9</u>	1.0	1.7	<u>5.9</u>
C-O-H	None	1.8	<u>3.2</u>	<u>8.5</u>	2.8	<u>4.6</u>	<u>12.0</u>
$RKC = 2.3$	Contour	0.9	1.6	<u>6.8</u>	1.4	<u>2.3</u>	<u>9.5</u>
	Strip	0.5	0.8	<u>3.4</u>	0.7	1.2	<u>4.8</u>
C-O-H-H-H	None	1.2	<u>2.1</u>	<u>5.6</u>	1.8	<u>3.0</u>	<u>7.8</u>
$RKC = 1.5$	Contour	0.6	1.1	<u>4.4</u>	0.9	1.5	<u>6.2</u>
	Strip	0.3	0.5	<u>2.2</u>	0.5	0.8	<u>3.1</u>
O-H-H-H	None	0.8	1.4	<u>3.7</u>	1.2	2.0	<u>5.2</u>
$RKC = 1.0$	Contour	0.4	0.7	<u>3.0</u>	0.6	1.0	<u>4.2</u>
	Strip	0.2	0.4	1.5	0.3	0.5	<u>2.1</u>

Soil Group 4; $K = 0.43$; $RK = 39$; Allowable Soil Loss 3 tons/acre

C-C-C	None	<u>6.6</u>	<u>11.5</u>	<u>30.3</u>	<u>9.8</u>	<u>16.4</u>	<u>42.6</u>
$RKC = 8.2$	Contour	<u>3.3</u>	<u>5.7</u>	<u>24.2</u>	<u>4.9</u>	<u>8.2</u>	<u>34.1</u>
C-C-O-H	None	<u>4.1</u>	<u>7.1</u>	<u>18.9</u>	<u>6.1</u>	<u>10.2</u>	<u>26.5</u>
$RKC = 5.1$	Contour	<u>2.0</u>	<u>3.6</u>	<u>15.1</u>	<u>3.1</u>	<u>5.1</u>	<u>21.2</u>
	Strip	1.0	1.8	<u>7.6</u>	1.5	2.6	<u>10.6</u>
C-O-H	None	2.8	<u>4.9</u>	<u>13.0</u>	<u>4.2</u>	<u>7.0</u>	<u>18.2</u>
$RKC = 3.5$	Contour	1.4	2.5	<u>10.4</u>	2.1	<u>3.5</u>	<u>14.6</u>
	Strip	0.7	1.2	<u>5.2</u>	1.1	1.8	<u>7.3</u>
C-O-H-H-H	None	1.8	<u>3.2</u>	<u>8.5</u>	2.8	<u>4.6</u>	<u>12.0</u>
$RKC = 2.3$	Contour	0.9	1.6	<u>6.8</u>	1.4	2.3	<u>9.6</u>
	Strip	0.5	0.8	<u>3.4</u>	0.7	1.2	<u>4.8</u>
O-H-H-H	None	1.3	2.2	<u>5.9</u>	1.9	<u>3.2</u>	<u>8.3</u>
$RKC = 1.6$	Contour	0.6	1.1	<u>4.7</u>	1.0	1.6	<u>6.7</u>
	Strip	0.3	0.6	2.4	0.5	0.8	<u>3.3</u>

TABLE 54 *(continued)*

Soil Group 5; K = 0.64; RK = 58; Allowable Soil Loss 4 tons/acre

C-C-C	None	9.7	16.9	44.8	14.5	24.2	62.9
RKC = 12.1	Contour	4.4	8.5	35.8	7.3	12.1	50.3
C-C-O-H	None	6.0	10.5	27.8	9.0	15.0	39.0
RKC = 7.5	Contour	3.0	5.3	22.2	4.5	7.5	31.2
	Strip	1.5	2.6	19.1	2.3	3.8	15.6
C-O-H	None	4.2	7.3	19.2	6.2	10.4	27.0
RKC = 5.2	Contour	2.1	3.7	15.4	3.1	5.2	21.6
	Strip	1.1	1.8	7.7	1.6	2.6	10.8
C-O-H-H-H	None	2.8	4.9	13.0	4.2	7.0	18.2
RKC = 3.5	Contour	1.4	2.5	10.4	2.1	3.5	14.6
	Strip	0.7	1.2	5.2	1.1	1.8	7.3
O-H-H-H	None	1.8	3.2	8.5	2.8	4.6	12.0
RKC = 2.3	Contour	0.9	1.6	6.8	1.4	2.3	9.6
	Strip	0.5	0.8	3.4	0.7	1.2	4.8
H-H-H							
RKC = 0.23	None	0.2	0.3	0.9[a]	0.3	0.5	1.2[a]

[a] Predicted losses on 20% slopes would be about 1.6 times as great as those given for 15% slopes. [a] in the 15% slope columns indicate that predicted losses would be within tolerance limits on 20% slopes also.

Yield correlations

PURPOSE This exercise is designed to introduce students and others to yield measurements and yield correlations to the soils. Pages 119–129 of the textbook and the references should be studied in preparation for this sampling. This effort is suitable for research projects and for advanced-degree work as well as for student projects, and involves yield measurements on statistically sited plots from various crops located on different soil map units. As Figures 66 to 68 illustrate, crop growth and yields are directly determined by soil conditions —even within a single field where climate and management are uniform. The measurement of yields on different soils becomes especially critical when detailed soil maps are used as the base for the land-tax system, and where taxes are assessed from yield values for specific soils.

PROCEDURES Yields vary from year to year depending on soil, weather, and management. Long-term yields are commonly estimated for crops grown in a survey area (see the textbook), but more precise measurements are needed on which to base more exact estimates. The thesis by Robinette (1975) excellently describes sampling procedures used for corn in Maryland. Sampled plots of two rows of corn each 20 feet long were randomly located and replicated four times in farmers' fields on the selected soil map units; a total of 202 sites was harvested (see page 124 of the textbook). Yields on different soils within the same field showed clear results of different moisture and erosion. Yield measurements help considerably to adjust and fine-tune estimates of crop yields that have been made in the past (Table 55).

In Iowa, Henao (1976) summarized regression modeling of corn yield data from about 2,800 plot-years. Plots were 1/100 acre in size within quarter sections randomly located. Ear corn was harvested, weighed, and a 300-gram sample was taken to determine moisture content, and weight equivalents were calculated to convert ear corn to shelled corn at standard 15.5 percent moisture (Table 56). Some of the soil variables studied are listed in Table 57; soil properties directly related to yields were slope, biosequence, available water-holding capacity, erosion, organic carbon, drainage class, percent clay, bulk density, pH, available phosphorus, and available potassium.

In Tennessee, soil productivity studies have been accomplished through cooperation of different agencies since 1954; Bell and Springer (1958) have provided an excellent description of the sampling process. Cotton yields from small fields with gin receipts were relatively easily related to specific soil map units. For corn yields, conservationists weighed harvested ear corn from 75 feet of row at six locations within the soil map unit, according to a systematic procedure. The plant populations and the number of lodged plants were also determined at the time the corn was harvested. Moisture determinations were made on shelled grain samples from 18 ears (three from each harvest plot). Yields were finally adjusted to a standard 15.5 percent moisture content (Table 56).

CORN YIELDS To sample corn yields within soil map units in a field, select an area with relatively uniform soil conditions in each soil map unit. The central part of a soil map unit should be selected, and areas of disturbance or abnormal treatment should be avoided. Divide the sampling area into 25 equal numbered parts and select five at random (from a random numbers table or by drawing numbers from a hat). From each area provide the information requested in the form shown in Table 58; the form can be modified to fit specific conditions in each state or region. The information on management can be obtained from the farmer.

An Ap horizon (topsoil) sample should be collected from each sampled plot; about 20 soil cores within the sampled plot should be mixed together for a soil sample of about 1 pound (1 pint or 1 liter volume size). A subsoil sample should also be collected from a small pit dug in the center of each plot. Any plots located on inclusions of different soils within the soil map units should be discarded, and another random sample should be chosen. Thus, a gravel spot within a map unit of clayey soil should not be sampled. The topsoil and subsoil samples collected from each plot should be submitted to a soil test laboratory for as complete analyses as possible. Soil test results can then be correlated with each plot yield.

For sampling each plot, measure the row width and count the number of plants. Row width is determined by averaging the width of 10 rows; average distance between rows is obtained by measuring the distance covered by 11 corn rows and dividing the distance by 10. Length of the two rows harvested can be varied to give a 1/250-acre sample (1/50-acre sample for five plots), according to the following adjustments for row spacing:

	Inches between Rows					
	40	38	36	34	32	30
Single row length to be harvested	52'4"	55'	58'	61'8"	65'4"	69'8"

Corn ears are harvested by hand from each 1/250-acre plot from two rows and weighed (a dairy scale works well for the weighings). About 300 grams of corn kernels is shelled from several ears and placed in an airtight metal drying can; the can is weighed, the lid is removed, and the corn kernels are dried in an oven for 48 hours at 65°C. After drying, the corn and can are weighed again and the moisture content and dry weight are calculated according to the equation

$$\% \text{ moisture} = \frac{\text{wet weight (g)} - \text{dry weight (g)}}{\text{wet weight (g)}} \times 100$$

$$\% \text{ dry weight} = 100 - \% \text{ moisture}$$

Remember that percent moisture of plant samples is based on wet weight; percentage moisture of soil samples is based on dry weight. Other electronic and radiometric methods can also be used to determine moisture in plant material and soil samples; all these methods require more expensive equipment.

The moist ear corn weight per bushel can be obtained from Table 56 where ear corn weights at different moisture percentages have been converted to weight per bushel at standard 15.5 percent moisture. The number of bushels can be calculated from the weighed yield for 1/250 acre by the equation

$$\text{plot yield (bu)} = \frac{\text{wet sample weight (lb)}}{\text{conversion weight}}$$
(from Table 56 in lb/bu)

A plot yield of 80 pounds wet sample weight (17 percent moisture), for example, would indicate

$$\frac{80 \text{ lb}}{70 \text{ lb/bu}} = 1.1 \text{ bu for 1/250 acre}$$

For conversion to a full acre, multiply 1.1 × 250 to give 275 bushels per acre for that plot measurement.

If the number of ears are counted in a plot and the average weight of ears in the sample plot are determined and converted to 15.5 percent moisture, the number of ears harvested can be multiplied by 250 to convert to a full acre, and the yield in bushels per acre represented by that plot can be read from Table 59. Thus, 32,000 corn ears at 0.6 pound each at 15.5 percent moisture represent a yield of 274 bushels per acre, or the same yield can be obtained from 24,000 ears at 0.8 pound each.

If the corn crop is to be sampled for silage yield, a few of the entire stalks from a plot must be cut, weighed, and an appropriate sample of several stalks dried for moisture determination. Laboratory analyses can be made of silage samples for total digestible nutrients and other determinations. The principles of yield sampling illustrated in these descriptions and references can be applied to any crop with the appropriate revisions.

The examples given in this exercise, of course, are not absolute, but the 1/250-acre sampling has been used by R. H. Rust and others when sampling yields on soils in Minnesota. Other sampling programs need to be flexible in formulating yield collection programs within states, regions, and counties. Where soils are more uniform, smaller plots and fewer replications are necessary. The nature of yield measurements depends to a great extent on the particular soil landscapes being sampled and the size of soil map units in relation to size of fields. A publication by Olson (1979) gives details about yield sampling programs that are under way in a number of places.

SHORTCUTS For characterization of performance of soil map units, many techniques can be used to measure yields. Although yield measurements are generally laborious, many shortcuts can be devised to speed up the process of making measurements. Table 59 illustrates a compilation of average weight of corn ears in comparison to ears harvested per acre—giving directly the resultant yields in average number of bushels per acre.

Thus, in determining yields quickly on different soil map units, ear counts can be easily made for plots in cornfields of 1/100 acre or so—and weighings can be made on samplings of random-sampled ears within the counted plots. Then the estimated bushels per acre can be made directly from Table 59. The data in Table 59, of course, are based on conversion of measurements of ear corn at variable moisture contents to shelled corn at standard 15.5 percent moisture, so that the crop moisture must be determined on a small sample before the conversions can be made. Numerous devices are now on the market to make moisture determinations of crop samples very easily and quickly.

Numerous gadgets have also been devised to make yield measurements and estimates easier. An electronic capacitance meter to estimate yields of forage *in situ* was developed and tested at the University of Nebraska (Burzlaff et al., 1973). Correlation coefficients (r) for alfalfa measurements were 0.80 for the first cutting and 0.73 for the second cutting as compared to conventional harvesting procedures. In Hawaii, direct visual estimates of average foliage height on a 28 X 28 cm sampling area were made with a device that utilized a plastic fresnel lens (Whitney, 1974). On grazed grasses an average of 7.5 readings/minute could be made. Number of samples required to estimate the mean forage height was eight to nine samples per plot for estimates ±2 cm and four per plot for ±3 cm; average foliage height explained 94 percent of the variation in dry matter yields. In Missouri, a disk meter was used to measure dry matter yields of forages (Bransby et al., 1977). Fifty paired observations of meter readings and yield measurements were made to establish the regression relationship; correlation coefficients (r) for bulk-height and dry matter yield ranged from 0.79 to 0.94, and all were significant at the 1 percent level of probability. For weighings, vegetation from a plot 9.6 ft² weighed in grams and multiplied by 10 gives a value equivalent to pounds per acre (see page 125 of the textbook).

These methods of crop sampling for yield correlations to soils, of course, are only provided as examples. Each person sampling yields on soils must be aware of the local crop, soil, environmental, and management conditions. Adaptations must be made, and constant checking must be done to ensure statistical validity of the data. When shortcuts are used, checks must be made on moisture determinations, reproducability of results, replications necessary, and so on. Throughout this exercise, the soil map must be checked to ensure that it is accurate to the scale at which the crop is being sampled. In this fashion (in systematic long-term programs), yield correlations will be considerably improved for predictive purposes in the future.

REFERENCES

Bell, F. F. and D. K. Springer. 1958. Cooperative studies in soil productivity and management in Tennessee. Journal of Soil and Water Conservation 13:156–157.

Barnsby, D. I., A. G. Matches, and G. F. Krause. 1977. Disk meter for rapid estimation of herbage yield in grazing trials. Agronomy Journal 69:393–396.

Burzlaff, D. F., G. L. Ham, and W. R. Kehr. 1973. *In situ* estimation of alfalfa forage yields. Agronomy Journal 65:644–646.

Garrard, H. L. 1979. Springtime diagnostic approaches for corn. Crops and Soils Magazine 31:12–16.

Henao, J. 1976. Soil variables for regressing Iowa corn yields on soil, management, and climatic variables. Ph.D. thesis, Dept. of Agronomy, Iowa State University, Ames, IA. 315 pages.

Olson, G. W. 1979. The "state-of-the-art" in linking soil fertility work to the soil survey (unabridged report). Soil Conservation Service, U.S. Department of Agriculture, Washington DC/Dept. of Agronomy, Cornell University, Ithaca, NY. Reproduced as Cornell Agronomy Mimeo 79-29. 143 pages.

Robinette, C. E. 1975. Corn yield study on selected Maryland soil series. M.S. thesis, Agronomy Dept., University of Maryland, College Park, Md. 192 pages.

Whitney, A. S. 1974. Measurement of foliage height and its relationships to yields of two tropical forage grasses. Agronomy Journal 66:334–336.

FIGURE 66/*Variable corn growth across a soil map unit boundary in Lewis County in northern New York State. The entire field has the same climate and management, but yields are vastly different in various places within the field due to the soils. Yield measurements show the specific responses of the crop to the soils. Obviously, a different cropping system is needed for the soil map unit in the foreground.*

FIGURE 67/*Variable legume mixture in a hayfield in Lewis County in northern New York State. Alfalfa on somewhat poorly drained soils and poorly drained soils has a high mortality rate due to frost heaving that disrupts the plant roots. Well-drained and moderately well-drained soils, in contrast, can support good legume growth under proper management. Different management is needed for the different soils in this field, and the field boundaries should be adjusted in accord with the soil map unit boundary.*

FIGURE 68/*Variable growth of corn and pasture plants in Lewis County in northern New York State. Yield measurements in the different soil areas would quantify the benefits to be derived by readjusting the field boundaries and management practices to fit the soil map units. Soils in the right part of the photo are the most productive, and should be farmed more intensively. Soils in the central part of the photo need drainage and other management corrections, and probably would be most profitable in the future in carefully managed pasture or hayland.*

TABLE 55/*Measured yields and estimated yields (bu/acre) of corn on several nearly level soils of Maryland (adapted from Robinette, 1975), and suggested changes for predicted yields.*

Soil	Measured Yield (bu/acre)	Estimated Yield (bu/acre)	Change
Beltsville	115	95	+20
Fallsington	105	120	−15
Lakeland	65	90	−25
Matapeake	140	140	0
Othello	100	115	−15
Sassafrass	115	130	−15
Chester	135	135	0
Calvin	115	82	+33
Gilpin	110	90	+20
Hagerstown	140	135	+5
Penn	125	90	+35

TABLE 56/*Weight equivalents for ear corn with a given initial moisture content (adapted from Robinette, 1975). Ear corn weight (lb/bu) indicates the weight of ear corn at an initial moisture content required to shell one bushel (56 lb) of grain adjusted to 15.5 percent moisture.*

Initial Moisture (%)	Ear Corn Weight (lb/bu)	Initial Moisture (%)	Ear Corn Weight (lb/bu)
12.0	65.0	22.0	76.8
12.5	65.5	22.5	77.5
13.0	66.0	23.0	78.3
13.5	66.3	23.5	79.0
14.0	66.8	24.0	79.7
14.5	67.3	24.5	80.5
15.0	67.9	25.0	81.2
15.5	68.4	25.5	81.9
16.0	68.9	26.0	82.7
16.5	69.5	26.5	83.5
17.0	70.0	27.0	84.2
17.5	70.7	27.5	84.9
18.0	71.3	28.0	85.7
18.5	71.8	28.5	86.3
19.0	72.6	29.0	86.9
19.5	73.2	29.5	87.8
20.0	73.9	30.0	88.5
20.5	74.5	30.5	89.2
21.0	75.3	31.0	89.9
21.5	76.1	31.5	90.6

TABLE 57/*Primary soil variables used for regression analyses (adapted from Henao, 1976).*

Soil map unit number, location
Slope configuration
Erosion class
Depth of A horizon (A1 + A2 + A3) in inches
Estimated percent organic carbon of 0-7-inch layer
Estimated percent organic carbon of 7-20-inch layers
Color value of 0-7-inch layer
Color chroma of 0-7-inch layer
Color value of 7-20-inch layers
Color chroma of 7-20-inch layers
Natural internal drainage
Subsoil permeability
Percent clay in plow layer
Maximum percent clay in subsoil
Average percent clay to 60 inches
Depth to midpoint of horizon with maximum percent clay
Subsoil group rating for crop growth
Biosequence
Geomorphic location on landscape
Bulk density at 15-30 inches
Bulk density at 30-40 inches
Subsoil structure
Parent material grouping
Depth to underlying contrasting strata

TABLE 57 *(continued)*
Minimum pH in subsoil
Depth to midpoint of minimum pH layer in subsoil
Thickness of minimum pH zone
Depth to top of carbonate horizon
pH of 10–20-inch zone
pH of 30–42-inch zone
pH of 42–60-inch zone
Available P of 10–20-inch zone
Available P of 20–42-inch zone
Available K of 12–24-inch zone

TABLE 58/*Form for recording information for measuring corn yields on soil map units (adapted from Robinette, 1975).*

Soil map unit name: _____ Date: _____

Map symbol: _____ Site No.: _____

Slope: _____ Soil Map No.: _____

Slope aspect: _____ Region: _____

Erosion: _____ County: _____

Comments: _____

_____

Operator's name: _____

Address: _____

Farm location: _____

History of field: _____

Rotation-cropping system: _____

Conservation and management practices: _____

	1983	1982	1981	1980	1979

Yield: _____

Fertilizer: _____

Lime: _____

Present crop: _____ Plant population: at planting _____

Variety: _____ at harvest _____

Planting date: _____ Row width: _____

Yield goal: _____ Length of growing season: _____

TABLE 58 *(continued)*

Fertilizer applied: _____   Harvesting date: _____

Soil management and treatment: Applied Nutrient Summary

Plowing: _____

	N	P₂O₅	K₂O	Mg

Disking: _____

Cultivating: _____   (lb)

Soil test results: _____

Lime: _____

pH	Mg	P₂O₅	K₂O	Ca

Herbicides: _____

Actual yield: _____

Topsoil: _____   Experimental: _____

Subsoil: _____   Operator: _____

TABLE 59/*Relationship between ear sizes and numbers and corn yield (average number of bushels per acre) (adapted from Garrard, 1979).*

Ears Harvested per Acre	Average Weight/Ear (lb) (15.5% moisture)					
	0.5 lb	0.6 lb	0.7 lb	0.8 lb	0.9 lb	1.0 lb
10,000	71	86	100	114	129	143
12,000	86	103	120	137	154	171
14,000	100	120	140	160	180	200
16,000	114	137	160	180	206	229
18,000	129	154	180	206	232	257
20,000	143	171	200	228	257	286
22,000	157	189	220	252	283	314
24,000	171	206	240	274	309	343
26,000	186	223	260	297	334	371
28,000	200	240	280	320	360	400
30,000	214	267	300	342	386	429
32,000	228	274	320	360	412	458
36,000	257	309	360	411	441	514
40,000	286	343	400	457	514	572
42,000	300	360	420	480	540	600

Farm planning

PURPOSE Practically all farms and fields, even small ones, have several different kinds of soils (Figure 69). This exercise is designed to introduce students and others to concepts of managing soils in accord with their capabilities. Often, tremendous increases in efficiency of farm operations can be achieved just by making a few adjustments in field boundaries. Wet, steep, stony, and other contrasting portions of fields represent different soil map units—and should not be included in fields that have other management requirements. Ideally, each soil has a "best use" according to the needs and desires of the farm operator and society as a whole. Management improvements like erosion control, irrigation, and drainage can be developed most efficiently if each soil map unit is used in harmony with the environment. Pages 98–129 of the textbook and the references cited in this section are especially important for farm planning.

PROCEDURE Each student should make a farm plan for his or her study area or project area, or for some other area in which he or she may be interested. More than 1 million farms and ranches in the United States have soil and water conservation plans prepared by the Soil Conservation Service, through the activities of nearly 3,000 Soil and Water Conservation Districts covering about 93 percent of the land in farms and ranches (SCS, 1973). An excellent brochure (SCS, 1973) has been prepared to answer the question "What is a farm conservation plan?"; each teacher should procure copies of this brochure to give to the students. An excellent slide set (SCS, 1969) is also available; this slide set should be procured by the teacher and shown to the students before they begin their activities in farm planning. The slide set introduces the concept of farm planning according to land capability classification and gives examples of landforms and management systems on each of the land capability classes in different regions of the United States. The slide set also illustrates excellent use of photographs that will be helpful to students preparing reports about soils.

The most important aspects of farm planning are in the attitudes and needs of the farm operator. Each student should consult the farmer managing the area in which the student is interested. Probably the landowner or operator already has a soil and water conservation plan that has been previously prepared; often these older plans need updating. Many farm plans have been made on unpublished soil maps that need to be updated where new published soil surveys are now available. Teachers may find it worthwhile to invite a progressive farmer or rancher to give a guest lecture to the class about experiences in farm planning, and to take a field trip to a specific farm to see conservation practices of modern management problems and solutions. This exercise provides an excellent opportunity for students to research all relevant sources of information about soil management for cropping (Olson, 1982).

EXAMPLE The farm planning brochure (SCS, 1973) includes an excellent example format to illustrate a farm conservation plan. This exercise description includes Figures 70 to 72 from the brochure (SCS, 1973) to show a soil and capability map (Figure 70), land use on a farm before a conservation plan (Figure 71), and land use on a farm after conservation farm planning (Figure 72). Table 60 (SCS, 1973) records some of the operator's decisions in farm planning.

As an example (SCS, 1973), a cotton and beef cattle farm in the southeastern United States is shown in Figures 70 to 72. Before 1940 the farm had tenants and sharecroppers who raised cotton with mules in a row crop system that promoted excessive erosion to the point where the productivity of the farm declined. After 1960 the farm had been converted into a more efficient operation where row crops, hay, and pasture could be farmed with power equipment. Beef cattle used hay and pasture on land that had been previously severely eroded from growing cotton on steep slopes. Bottom land is currently used for cotton where erosion is not so great a problem. A hunting area for doves was added as an income-raising recreation enterprise. The soils on this farm (Figure 70) and their capability (Figure 70) are:

IIe3   40B2—Grenada silt loam. Nearly level or gently sloping land with slight or moderate erosion. Soil is moderately deep and moderately well drained and has a pan at a depth of about 20 inches.

IIw2   2A—Collins silt loam. Silty soil in slightly wet bottom land that may overflow. May be either acid or alkaline.

IIw3   4A—Falaya silt loam. Somewhat poorly drained bottom land subject to moderate flooding.

IIw4   43A1, 43B1, 43B2—Calloway silt loam. Nearly level to gently sloping land with slight to moderate erosion. Shallow, cold, and somewhat poorly drained soil.

Soils (SCS, 1973) in Class II have some natural condition that limits the kinds of plants they can produce or that calls for some easily applied conservation practice when they are cultivated. They are suited for use as cropland, grassland, woodland, wildlife land, or recreation land.

Soils (SCS, 1973) in Class III have more serious or more numerous limitations than those in Class II (Figure 70). Thus, they are more restricted in the crops they can produce or, when cultivated, call for conservation practices more difficult to install or keep working efficiently. They are suited for use as cropland, grassland, woodland, wildlife land, or recreation land.

IIIe1   31C3—Loring silt loam; 39C2—Memphis and Natchez silt loams. Gently to moderately sloping land with moderate to severe erosion. Soils are deep and well drained.

IIIe4   40B3—Grenada silt loam. Gently sloping, severely eroded land. Soil is shallow or moderately deep and moderately well drained and has a pan at a depth of about 20 inches.

IIIe5   40C2—Grenada silt loam. Moderately sloping land with slight to moderate erosion. Soil is moderately deep and moderately well drained and has a pan at a depth of about 20 inches.

IIIw3   23A—Henry silt loam. Nearly level, wet, cold land. Soil is shallow and poorly drained.

Soils (SCS, 1973) in Class IV have very severe limitations that require very careful management. They are suitable for occasional but not regular cultivation and for grassland, woodland, wildlife land, or recreation land.

IVe1   31D3—Loring silt loam. Steep land with moderate to severe erosion. Soil is deep and well drained.

IVe4   40C3—Grenada silt loam. Moderately sloping land with severe erosion. Soil is moderately well drained and has a silty surface and subsoil and a pan at a depth of about 20 inches.

Soils (SCS, 1973) in Class VI have severe limitations that make them generally unsuited for cultivation and that restricted their use to pasture, range, woodland, recreation, or wildlife food and cover.

VIe1   39E3—Memphis and Natchez silt loams. Steep to very steep land with moderate to severe erosion. Soils are deep and well drained with silt loam surfaces and silty clay loam subsoils.

VIe2   40D3—Grenada silt loam. Steep land with severe erosion. Soil is moderately deep to a pan and is moderately well drained.

Soils (SCS, 1973) in Class VIII have very severe limitations that make them unsuited for cultivation and that restrict their use to pasture, range, woodland, recreation, or wildlife food and cover with careful management.

VIIe3   60—Gullied land. Moderately sloping to very steep land, severely gullied. Underlying materials range from silt loams to gravels and clays.

For woodlands (SCS, 1973), loblolly pine is best suited for planting on Grenada silt loam (40B2, 40B3, 40C2, 40C3, 40D3) and can produce about 230 board-feet per acre per year. Because of the pan in the soil some windthrow is likely. Commercial hardwoods such as cherry-bark oak and sweetgum make satisfactory growth on uneroded areas. Cottonwood, cherry-bark oak, willow oak, and sweetgum on Collins silt loam (2A) produce about 500 board-feet per acre per year. Cherry-bark oak or sweetgum will produce about 400 board-feet on Memphis and Natchez silt loams (30E3), but steep slopes cause some management problems. On the gullied land (60), loblolly pine will give the best erosion control, watershed protection, and growth.

RESULTS Results of farm planning for 1940–1960 are illustrated in Figures 70 to 72 and Table

60. Cropland was shifted from sloping eroded fields to nearly level uneroded soils. Field boundaries were adjusted to conform more closely to soil map units. Costs were reduced, and profits increased. Increases were achieved in pastureland, hayland, wildlife land, and woodland. Idle land was reduced. Before 1940, a few scrub cattle subsisted on abandoned cropland. After 1960, more than 100 beef cattle were fed on pasture, hay, and silage.

Before 1940, soil erosion was as high as 225 tons per acre per year in places (SCS, 1973). After 1960, rates were reduced to 3 tons per acre per year to 1 ton per acre per year. Soil fertility is being restored. Bottomlands no longer scour, improved channels now hold most of the runoff without flooding, and special outlets protect channels entering the main drainageways.

Remember that farm planning is not accomplished overnight. On this farm, more than 20 years of farm planning and implementation of those plans was necessary to reduce erosion and begin to restore the fertility of the soil. Societal inputs were necessary in providing technical and financial assistance to the farmer. Everyone benefits, however, in investment to improve land use and management of soil map units through farm planning.

To complete this exercise, each student should submit a report on a farm planning effort similar to the example described here. The report can be brief or lengthy and elaborate, depending on the time and priority given to farm planning in relation to the rest of the course. Increasingly, environmental and agricultural professionals will be involved with farm planning. In New York State, for example, legislation has been recently passed that will require soil and water conservation plans for all properties of 20-acre size or larger. Landowners and others involved with land use would be well advised to learn as much as they can about farm and land-use planning for improving soils and the environment in the future.

REFERENCES

Olson, G. W. (Editor). 1982. 1982 Cornell recommends for field crops. New York State College of Agriculture and Life Sciences, Cornell University, Ithaca, NY. 56 pages.

SCS. 1969. Know your land: Narrative guide, with photographs and captions, and color slide set. Soil Conservation Service, U.S. Dept. of Agriculture, U.S. Government Printing Office, Washington, DC. 12 pages.

SCS. 1973 (revised). What is a farm conservation plan? PA-629, Soil Conservation Service, U.S. Dept. of Agriculture, U.S. Government Printing Office, Washington, DC. 8 pages.

FIGURE 69/*Landscape of a typical farm in New York State. Most farms have nearly level soils and steep soils, droughty soils and wet soils, coarse-textured soils and fine-textured soils, erosive soils and soils not subject to severe erosion, and many other highly contrasting soil map units. The challenge is to match the management of the farm to the capabilities or "best use" of each soil map unit.*

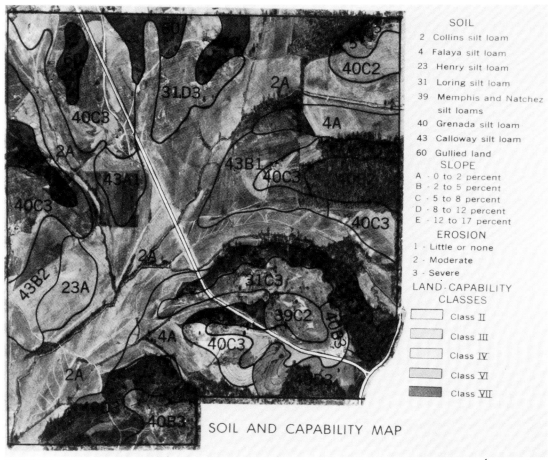

FIGURE 70/*Soil and capability map for example farm in the southeastern United States (adapted from SCS, 1973). The soil map is the base for farm planning, and the land capability classes are groupings of soil map units for management purposes.*

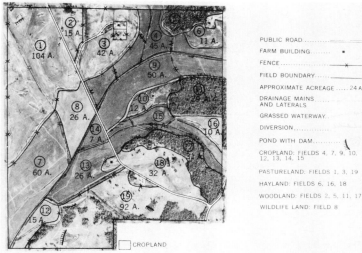

PUBLIC ROAD..........
FARM BUILDING..........■
FENCE..........—×—
FIELD BOUNDARY.........._____
APPROXIMATE ACREAGE.....24 A.
DRAINAGE MAINS
AND LATERALS..........
GRASSED WATERWAY..........
DIVERSION..........
POND WITH DAM..........
CROPLAND: FIELDS 4, 7, 9, 10,
12, 13, 14, 15
PASTURELAND: FIELDS 1, 3, 19
HAYLAND: FIELDS 6, 16, 18
WOODLAND: FIELDS 2, 5, 11, 17
WILDLIFE LAND: FIELD 8

FIGURE 71/*Land use before 1940 on example farm in the southeastern United States (adapted from SCS, 1973). Compare the land use with the soil and capability map in Figure 70. Notice that numerous areas of steep slopes are farmed to cause severe erosion, and that farming of small patches is inefficient.*

FIGURE 72/*Land use after 1960 on example farm in the southeastern United States (adapted from SCS, 1973). Compare the land use (cropland) with Figure 71, and compare land-use relationships with the soils mapped in Figure 70. After 1960 the farm had become much more efficient and productive through farm planning.*

TABLE 60/*Record of operator's decisions in farm planning (adapted from SCS, 1973).*

Field	Acreage	Use and Conservation Treatment
1, 3, 19	238	Pastureland—Construct diversions and three shaped grassed waterways. Smooth gullies. Build four ponds and fences. Lime, fertilize, and establish bermudagrass and annual lespedeza. Control weeds with chemicals. Relime and refertilize for moderate yields. Rotate grazing and prevent overuse. Stock the large pond in Field 19 with largemouth bass and bluegill. Fertilizer water for heavy fish production. Install safety equipment at pond.
4, 7, 9, 12, 13, 14, 15	236	Cropland—Construct drainage channel; fill old channel sections with spoil from new channel; build one levee road with spoil; build drainage laterals and field ditches and six shaped grassed waterways. Install a grade-stabilization structure (drop inlet) at each location where a lateral enters the drainage channel. Grow annual crops of corn, cotton, and sorghum silage. Fertilize for profitable yields, shred corn and cotton stalks for mulch, and arrange rows for good drainage. Irrigate crops with sprinkler system from nearby upstream irrigation reservoirs. Leave grass unmowed along channel for rabbits.
10	12	Cropland—Grow truck crops each year. Construct diversions. Cultivate on the contour. Use barnyard manure and fertilize liberally.
6, 16, 18	53	Hayland—Smooth gullies in Field 18. Lime and fertilize for profitable yields. Establish bermudagrass and annual lespedeza. Refertilize and relime to maintain yields.
8	26	Wildlife land—Each year grow browntop millet in cultivated rows for doves and allow hunting in season for income. Plant about 70 days before hunting season. Fertilize with about 500 pounds of 6-12-6 to the acre.
2	15	Woodland—Establish full stand of loblolly pine. Exclude livestock. Keep out fire.
5, 11, 17	73	Woodland—Manage existing woodland for such hardwoods as white oak, cherry-bark oak, sweetgum, and yellow poplar. Make intermediate cutting every tenth year beginning in 1963. Keep out livestock and wildfire. Leave an average of three trees per acre of hickory, mulberry, walnut, or elm for squirrels.
	19	Miscellaneous—Establish and maintain ground cover as necessary to control erosion.

Community planning

PURPOSE For planning of urban communities, this exercise is designed to parallel the preceding exercise on farm planning. Increasingly, recognition is being given to the fact that soils are fully as important to community planning as to farm planning. Indeed, planning on a local, regional, and national scale must consider interrelationships of rural and urban communities in natural entities of drainage basins and soil regions—if the planning is to be efficient and effective. Wetness, shrink–swell, erosion, runoff, slope, texture, depth to bedrock, instability, stoniness, and many other soil characteristics are vital components to be considered in community planning. Behavior of roadbeds, foundations, waste disposal operations, landscape design, community aesthetics, architecture, human health and well-being, industrial plant maintenance, extent of parklands, efficiency of population-sustaining areas, energy necessary for local production versus costs of imports, and so on, are all highly dependent on the characteristics of the soils. As for agriculture, each soil has a "best use" in a community according to the needs and desires of the individuals, the community, and society as a whole. Location and maintenance of buildings, roads, parks, shopping areas, industries, schools, waste facilities, farms, forests, and other land uses should be in harmony with the natural environment and the soils. If constructions are improperly located, the results can be disastrous; many examples can be cited of past mistakes where soil information was not adequately considered. For this exercise in community planning, students should study pages 53-97 and 130-160 of the textbook and the references cited in this section.

PROCEDURE Each student should make a plan for a community—real or hypothetical, large or small. Possibly some students are already doing community planning as a part of the overall projects for the course as suggested in Tables 29 to 31. An excellent brochure (SCS, 1973) has been prepared as a guide entitled "A conservation plan...for a developing area"; each teacher should procure copies of this brochure to give to the students. Important considerations must be given to "comprehensive planning" where all of the

resources of the soils and landscapes are considered to meet the total needs of the community. The greatest opportunities, of course, are presented when totally new cities are built where none existed before—as in the case of Canberra, Australia; Brasilia, Brazil; Chandigahr, India; Columbia, Maryland; and other "new cities." More often, planning projects must also consider older communities that may have been built with virtually no planning. Worthwhile efforts for students and teachers engaged in planning activities are to collect newspaper articles relating to problems caused in their community as a result of lack of adequate considerations of the soils. Students can then suggest soil improvements, modifications, and plans for projects to overcome the soil limitations and improve the community for people in the future.

Many references are available to assist in community planning; these references should be acquired by the teacher and kept in a library where they are readily available to students. A book edited by Bartelli et al. (1966) gives a good review of soil surveys and land-use planning in different parts of the United States. A book edited by Beatty et al. (1979) deals with planning the uses and management of land, although not all of the chapters deal exclusively with soil surveys. A book by McHarg (1969) gives the landscape architecture and comprehensive planning perspective, with excellent examples of overlay map techniques. A bulletin by Olson et al. (1969) presents an example of planning with soils information for the five-county area around Syracuse. Simonson (1974) edited a book on nonagricultural applications of soil surveys containing examples of studies from the Netherlands, Canada, Australia, and New Zealand. Davidson (1980) and McRae and Burnham (1981) review techniques of land-use planning and land evaluation from the British point of view. All these references will be of considerable assistance to students in doing projects involving use of soil surveys for community planning.

EXAMPLE The brochure (SCS, 1973) describing "a conservation plan...for a developing area" includes an excellent example format to illustrate community planning. Figure 73 (SCS, 1973) illustrates a soil map made at a large scale especially for community development. Table 61 (SCS, 1973) gives brief descriptions and soil interpretations for the soil map units in Figure 73. Obviously, houses and other buildings should not be put on the Es Eel soils of the floodplain. The HeF Hennepin and MnC and D Miami soils on steep slopes have severe limitations. In contrast, MnB2 Miami soils have slight limitations for most uses, and the other soils have limitations that can be modified or improved with careful design and management. For community planning and development, Figure 74 (SCS, 1973) outlines how planning and development can be fit to the natural landscape and the soils. Large buildings can be put on the MnB2 soils that have slight limitations, if erosion is carefully controlled. Houses can be put on the CrA and MnB2 soil map units with careful design and maintenance and preservation of the natural vegetation as much as is possible; the CrA soil map units do have wetness problems requiring drainage for house foundations, basements, septic tanks, and streets. Steep slopes and floodplains are carefully protected for aesthetics and drainage, but can be developed for parklands and open space. During construction, Table 62 outlines specific measures that can be taken to reduce sediment losses and achieve most efficient and effective landscape design. Similarly, students should select an area likely to be developed near their place of residence, investigate all they can about the soils and proposed developments, and submit a community plan for the area that considers the soils and the environment. Project reports should be graded by giving major considerations to initiative, imagination, neatness, thoroughness, organization, use of photographs, and applicability of results of the findings of the student research.

COLUMBIA Many case studies exist which teachers and students and others can use to see the effects of soil survey interpretations on community planning. One of the best illustrations of good planning is the new city of Columbia, Maryland, between Baltimore and Washington, D.C. Columbia, started in the 1960s, "was laid out to harmonize with the environment and serve as a model for an urban ecology where nature and humanity are not estranged" (Shabecoff, 1982). Much has been written about this new city and other new towns (Breckenfeld, 1971; Eichler and Kaplan, 1967).

Columbia (in 1982) had about 55,000 people in an area of 15,000 acres, where more than 800,000 trees and shrubs have been planted in the last 20 years. Floodplains were preserved and developed for parks and picnic areas (Figure 75). Streets were laid out along topographic contours and soil boundaries (Figure 76), for more efficient and aesthetic results. A swamp area was turned into a lake rather than being drained at substantial cost for a building site (Figure 77). Steep slopes, expensive to level or to build on and more fragile, were left alone as much as was possible. Fields and woodlands were protected as natural meadows and forests, rather than being bulldozed. About 35 percent of the land in Columbia is retained as open space and parkland, and is carefully maintained and managed by the Columbia Park and Recreation Association. Houses, apartments, condominiums, schools, shopping centers, businesses, and industries are located in clusters on the soils most suitable for constructions. Power lines and utilities are placed underground to reduce visual pollution. A 1,200-acre wildlife preserve is available for all to enjoy, within the city. Throughout the city construction, soil erosion (Figure 78) was controlled and kept to a minimal amount.

Community planning in Columbia, of course, was not easy and automatic. A great deal of effort was expended to meet two fundamental goals:

1. To create a social and physical environment that would work for people, nourishing human growth.
2. To make a profit on the land development and sale, as a venture of private capital.

In planning Columbia, expert assistance was sought from soil scientists, conservationists, geologists, botanists, planners, engineers, economists, developers, bankers, architects, recreationists, educators, lawyers, politicians, sociologists, psychologists, public health officials, wildlife specialists, and many others. Street names and other place names, for example, were developed in accord with the history of various sites within the city and natural features and characteristics. Cluster development enabled preservation of open spaces. Schools and shopping facilities were placed near housing centers so that auto traffic could be minimized. Health and education facilities were innovative and efficient in use of land and space.

PRINCIPLES Principles of community planning in accord with soil characteristics (or lack of planning) can be demonstrated in nearly every community. In Sterling Forest, New York (near New York City), many houses have been built on soils that are stony, steep, or shallow to bedrock (Figure 79). Although these soils have severe limitations for houses, the developers were very careful to preserve the trees and site the houses with minimal soil and environmental disruptions. An important consideration is that better soils elsewhere were protected for other uses, and were not destroyed by urban sprawl of houses. Costs are greater to build on soils with severe limitations, but the costs must be evaluated on a long-term basis and not solely for short-term profit. Comprehensive planning of a large region or drainage basin enables many different soil map units to be evaluated, and decisions can be made on a more rational basis in the future considering the long-term needs of society.

Increasingly, soil maps are being made and used in urban areas. Figure 80 is a portion of a soil map of Anaheim, California, showing soils under urban constructions (houses, motels, restaurants, roads, etc.). This type of map is especially valuable to plan excavations for sewer and water and utility lines in the future, and serves as a guide for preliminary locations of buildings. Unfortunately, small isolated agricultural areas (Figure 81) are under great pressures, and probably will be lost to urban development in the future. With political support, however, agricultural districts can be created where agriculture can be protected on the best soils for future generations.

Comprehensive regional planning with soils maps is essential to allocate soil resources for the future. Urban sprawl not only occupies the best soils (Figure 81 and 82), but also creates pressures on other soils that have severe limitations (Figures 83 to 85). Thus, land abuse and misuse have severe consequences over broad areas, not just at a specific site in question. Comprehensive community planning, with detailed soils maps of large areas, will help to prevent some of the land abuse of the past. The future must be given higher priority than it has been given in the past, if peace, progress, and prosperity is to be achieved.

ASSIGNMENTS Assignments to students in community planning should include collection of newspaper articles illustrating relevant land-use problems and solutions relating to uses of soil maps. Newspapers are a reflection of the problems and potentials of our society. At the time of this writing (December 1982), for example, areas of HsC3 Hudson and related soils are being developed for urban constructions southeast of Ithaca, New York (see Figure 85). Considering the unstable nature of these soils in the Six Mile Creek area (Figures 2 and 3), it would seem unwise to build on these soils. In spite of the hazards, the Ithaca Town Planning Board decided unanimously "not to require an environmental impact statement" for a controversial housing project in the area (Mundell, 1982). Obviously, a great deal of education about the soils is needed by students, teachers, laypersons, planners, engineers, developers, politicians, and the general public. At Boynton school site in Ithaca, settling of alluvial floodplain soils has caused walls to crack and doors to stick, and will cost tens of thousands of dollars to correct (Scott, 1982). In Missouri, dioxin (toxic to human beings) has been deposited with soil flood sediments in urban communities (Biddle, 1982). At Hilton Head Island, South Carolina, excessive urban development has caused sewerage and sedimentation problems (Stuart, 1982). In Stowe, Vermont, legal controversy over a building permit for a motel in a 16-acre meadow caused the developer to start a pig farm in full view (and upwind) from the irate residents of that community (Anonymous, 1982). Truly, community planning involves emotions and politics as well as rational (or irrational) decisions about allocations of soils resources. Informed citizens (students, teachers, laypersons) are the only hope for better community planning for the future.

SOILS TOURS Students, teachers, and laypersons engaged in community planning should take full advantage of soils tours and other tours to see the full range of problems and potentials for improvements. Photographs included in this section, for example, were taken during tours in California, Maryland, and New York State (Dembroff et al., 1982; Olson, 1980; Singer, 1982). Teachers of soil survey interpretations and community planning should organize tours for students, and students should be inspired to take informal tours by themselves and with colleagues.

REFERENCES

Anonymous. 1982. With a motel permit stalled, developer

starts a pig farm. The New York Times. 19 Dec. Page 66.

Bartelli, L. J., A. A. Klingebiel, J. V. Baird, and M. R. Heddleson (Editors). 1966. Soil surveys and land use planning. American Society of Agronomy, Madison, WI. 196 pages.

Beatty, M. T., G. W. Petersen, and L. D. Swindale (Editors). 1979. Planning the uses and management of land. Monograph 21, American Society of Agronomy, Madison, WI. 1028 pages.

Biddle, W. 1982. Toxic chemicals imperil flooded town in Missouri. The New York Times. 16 Dec. Page A17.

Breckenfeld, G. 1971. Columbia and the new cities. Ives Washburn, Inc., New York. 332 pages.

Davidson, D. A. 1980. Soils and land use planning. Longman, New York. 129 pages.

Dembroff, G. R., E. A. Keller, D. L. Johnson, and T. K. Rockwell. 1982. The soil, geomorphology, and neotectonics of the Ventura river and central Ventura basin, California. Office of Earthquake Hazards, U.S. Geological Survey, Menlo Park, CA. 71 pages.

Eichler, E. P. and M. Kaplan. 1967. The community builders. University of California Press, Berkeley. 196 pages.

Gray, D. H., and A. T. Leiser. 1982. Biotechnical slope protection and erosion control. Van Nostrand Reinhold Company, New York. 271 pages.

McHarg, I. L. 1969. Design with nature. Doubleday Natural History Press, Garden City, NY. 197 pages.

McRae, S. G. and C. P. Burnham. 1981. Land evaluation. Claredon Press, Oxford, England. 239 pages.

Mundell, H. 1982. Commonland: Environmental impact statement not required, despite many requests. The Ithaca Journal. 6 Oct. Pages 1 and 11.

Olson, G. W. 1980. Tour of soils southeast of Ithaca in Tompkins County, New York. Fact Sheet—Soil Classification—Page 12:00. New York State College of Agriculture and Life Sciences, Cornell University, Ithaca, NY. 6-page leaflet with photos and map.

Olson, G. W., J. E. Witty, and R. L. Marshall. 1969. Soils and their use in the five-county area around Syracuse. Cornell Miscellaneous Bulletin 80, New York State College of Agriculture and Life Sciences, Cornell University, Ithaca, NY. 100 pages and map.

Scott, L. 1982. At Boynton (Middle School) settling has caused walls to crack and doors to stick. The Ithaca Journal. 18 Dec. Page 1.

SCS. 1973. A conservation plan...for a developing area. Program Aid No. 1029, Soil Conservation Service, U.S. Dept. of Agriculture, U.S. Government Printing Office, Washington, DC. 6 pages in folded leaflet.

Shabecoff, P. 1982. A city where man and nature coexist (Columbia, Maryland). The New York Times. 25 Nov. Page A18.

Simonson, R. W. (Editor). 1974. Non-agricultural applications of soil surveys. Elsevier Scientific Publishing Co., Amsterdam. 178 pages.

Singer, M. 1982. General soils and land use tour southeast of Anaheim, California, for Soil Science Society of America. Dept. of Land, Air, and Water, University of California, Davis. 29 pages.

Stuart, R. 1982. Hilton Head, seen as island paradise, is straining under big-city problems. The New York Times. 14 Dec. Page A18.

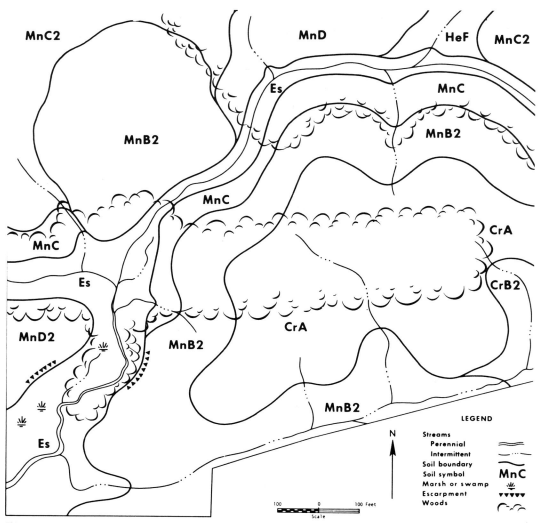

FIGURE 73/*High-intensity soil map at large scale for community planning and development (adapted from SCS, 1973).*

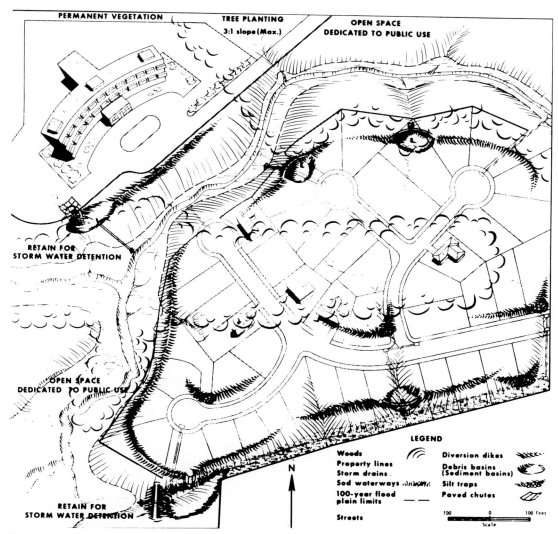

FIGURE 74/*Community development of soil map units shown in Figure 73, with the development in harmony with the soils and the environment (adapted from SCS, 1973).*

FIGURE 75/*Soils of floodplain preserved for parkland and recreation in Columbia, Maryland. Better-drained soils on the higher ground are developed for apartment houses, schools, and other heavy buildings. The floodplain preservation helps to maintain the natural runoff patterns, and lessens the damage when heavy rains occur.*

FIGURE 76/*Street layout along contour lines and soil boundaries in Columbia, Maryland. Steeper slopes are protected and relatively undisturbed, and aesthetic views are enhanced from the efficient contour placement of the housing sites. Traffic is also more efficient where road grades are minimized and drainage modifications are fewer.*

FIGURE 77/*Planned development of shore of artificial lake in Columbia, Maryland. Wet flooded soils were impounded with water, and sloping soils were stabilized with vegetation. The lakeshore is reserved as a "commons" that everyone can use, with walkways and parkland facilities. All people have access to the lakeshore, and no private property excludes joggers, walkers, or bicycles. Cluster development on the better soils enables more people to be put into smaller areas, but adjacent parklands are available as "open space" to everyone.*

FIGURE 78/*Soil erosion in Columbia, Maryland, during construction activities. Many of the soils in Columbia are highly erosive, so that slopes must be stabilized as quickly as possible (see Table 62). Sediment basins and mulching of soil surfaces have proved to be very beneficial in controlling erosion in Columbia.*

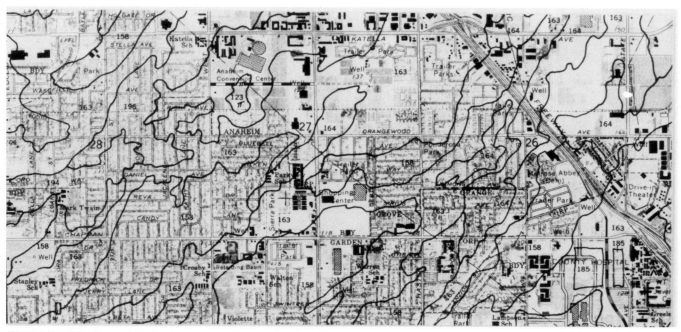

FIGURE 79/*House built on soils stony, shallow to bedrock, and steep with severe limitations. Soil modifications were carefully planned and implemented, so that better soils nearby could be used for farming and industrial development. Good community planning must be comprehensive, matching the characteristics of the soils and the environment to the needs of the community, the region, and the nation.*

FIGURE 80/*Soil map of urban area in Anaheim, California. Houses, motels, restaurants, and other constructions are rapidly displacing agricultural uses for the soils. Many of these soils have excellent capabilities for agricultural production, but that capability is lost when the soils are appropriated for urban uses. With comprehensive planning in accord with soil characteristics and potential, more rational decisions could be made about future implications of soil map unit allocations.*

FIGURE 81/*Sharp transition of agricultural and urban uses of soils in Anaheim, California. Farming cannot compete with profits from urban uses of the soils, and food production potential for the future will soon be lost in this area.*

FIGURE 82/*Urbanization in valley northwest of Ventura, California. The best soils for farming in this valley have been lost to urbanization. Future food production will have to come from other areas.*

FIGURE 83/*Steep, stony, shallow to bedrock soils being planted to avocado orchards near Palo Mission, California (southeast of Los Angeles near Camp Pendleton). The land costs thousands of dollars per acre, and the development costs tens of thousands of dollars per acre. Land pressures on these soils of extremely severe limitations are caused by the loss of better agricultural soils to urbanization elsewhere. Ultimately, society will be held accountable for the misappropriation of soil resources.*

FIGURE 84/*Housing constructions on hills north of Ventura, California. These soils are prone to earthquakes and landslides, but better soils have already been occupied by urban constructions. Biotechnical slope protection and erosion control (Gray and Leiser, 1982) will be critical to slope stabilization in these areas.*

FIGURE 85/*Landslide below Bethel Grove Bible Church southeast of Ithaca, New York (see Figures 2 and 3). When saturated, these soils become liquid and literally "flow." The HsC3 Hudson and related soils are questionable for development for urban constructions.*

TABLE 61/*Brief soil descriptions and interpretations for soil map units of development area shown in Figure 73 (adapted from SCS, 1973).*

Soil Descriptions

Symbol	Soil	Brief Description
CrA	Crosby silt loam, 0 to 2 percent slopes	Light-colored to moderately dark colored, deep, somewhat poorly drained, slowly permeable soils on nearly level to gently sloping areas in uplands. Developed from firm glacial till.
Es	Eel silt loam	Moderately dark colored, deep, moderately well drained, moderately permeable soils on nearly level areas in bottom lands. Developed from friable alluvium.
HeF	Hennepin soils, 18 to 35 percent slopes	Light-colored, deep, well-drained, moderately permeable soils on steep slopes in uplands. Developed from firm glacial till.
MnB2	Miami silt loam, 2 to 6 percent slopes, eroded	Light-colored, deep, well-drained, moderately permeable soils on gentle slopes to moderately steep slopes in uplands. Developed from firm glacial till.
MnC	Miami silt loam, 6 to 12 percent slopes	
MnC2	Miami silt loam, 6 to 12 percent slopes, eroded	
MnD	Miami silt loam, 12 to 18 percent slopes	
MnD2	Miami silt loam, 12 to 18 percent slopes, eroded	

Soil Interpretations

Symbol	Soil	Erosion Hazard[a]	Dwellings—Three Stories or Less: With Basement	Without Basement	Septic Tank Absorption Fields	Location of Roads and Streets	Parks or Nature Trails
CrA	Crosby	Slight	Severe; wetness	Moderate; wetness	Severe; wetness	Moderate; wetness	Moderate; wetness
CrB2	Crosby	Moderate; sheet erosion	Severe; wetness	Moderate; wetness	Severe; wetness	Moderate; wetness	Moderate; wetness
Es	Eel	Moderate; streambank erosion	Severe; floods	Severe; floods	Severe; floods	Severe; floods	Moderate; floods
HeF	Hennepin	Severe; sheet and gully erosion	Severe; slope	Severe; slope	Severe; slope	Severe; slope	Severe; slope
MnB2	Miami	Moderate; sheet erosion	Slight	Slight	Slight	Slight	Slight
MnC, MnC2	Miami	Severe; sheet and gully erosion	Moderate; slope	Moderate; slope	Moderate; slope	Moderate; slope	Moderate; slope
MnD, MnD2	Miami	Severe; sheet and gully erosion	Severe; slope	Severe; slope	Severe; slope	Severe; slope	Severe; slope

[a] Soil limitation classes: Soils rated as slight have few or no limitations for the use. Soils rated as moderate have limitations that reduce to some degree their desirability for the purpose being considered. They require some corrective measures. Soils rated as severe have unfavorable soil properties or features that severely restrict their use and desirability for the purpose. A severe rating does not mean that the soil cannot be used for a specific purpose because many of the problems can be corrected.

TABLE 62/*Measures recommended during construction to reduce erosion, stabilize exposed areas, and achieve efficient landscape design (adapted from SCS, 1973).*

1. Disturb only the areas needed for construction. At the present time, natural vegetation covers this area and there is little erosion. The streambed and streambanks are stable. The vegetation on the floodplain and adjacent slopes will contribute to the aesthetic and environmental quality of the development.
2. Remove only those trees, shrubs, and grasses that must be removed for construction; protect the rest to preserve their aesthetic and erosion-control values.
3. Stockpile topsoil and protect it with anchored straw mulch.
4. Install sediment basins and diversion dikes before disturbing the land that drains into them. Diversion dikes in the central part of the development may be constructed after streets are installed but before construction is started on the lots that drain into them.
5. Install streets, curbs, water mains, electric and telephone cables, storm drains, and sewers in advance of home construction.
6. Install erosion and sediment-control practices as indicated in the plan and according to soil conservation district standards and specifications. The practices are to be maintained in effective working condition during construction and until the drainage area has been permanently stabilized.
7. Temporarily stabilize each segment of graded or otherwise disturbed land, including the sediment-control devices not otherwise stabilized, by seeding and mulching or by mulching alone. As construction is completed, permanently stabilize each segment with perennial vegetation and structural measures. Both temporary and permanent stabilization practices are to be installed according to soil conservation district standards and specifications.
8. "Loose-pile" material that is excavated for home construction purposes. Keep it "loose-piled" until it is used for foundation backfill or until the lot is ready for final grading and permanent vegetation.
9. Stabilize each lot within four months after work starts on home construction.
10. Backfill, compact, seed, and mulch trenches within 15 days after they are opened.
11. Level diversion dikes, sediment basins, and silt traps after areas that drain into them are stabilized. Establish permanent vegetation on these areas. Sediment basins that are to be retained for stormwater detention may be seeded to permanent vegetation soon after they are built.
12. Discharge water from outlet structures at nonerosive velocities. Design debris basins as detention reservoirs so that peak runoff from the development area is no greater than that before the development was established.

Principles governing the applications of soil survey interpretations in the future

PURPOSE Soil potentials are index numbers assigned to specific soil map units in a survey area to indicate performance (e.g., yield) and costs (e.g., drainage, irrigation, maintenance) for using each soil for a specific purpose. The gathering of output and input data used in calculation of soil potentials requires interdisciplinary cooperation of soil scientists and soil conservationists with agronomists, assessors, contractors, economists, engineers, planners, and many others. Soil potentials have economic units or economic implications, because feasibility of development of soil areas is usually directly related to costs and returns (Beatty et al., 1979). This exercise is designed to acquaint students and others with soil potentials, and to provide practice in procedures to calculate ratings of soils in an area for their relative development potentials. Soil potentials are used (Soil Survey Staff, 1978):

1. To provide a common set of terms, applicable to all kinds of land uses, for rating the quality of a soil for a particular use relative to other soils in the area.
2. To identify the corrective measures needed to overcome soil limitations and the degree to which the measures are feasible and effective.
3. To enable local preparation of soil interpretations, using local criteria to meet local needs.
4. To provide information about soils that emphasizes feasibility of use rather than avoidance of problems.
5. To assemble in one place information on soils, corrective measures, and the relative costs of corrective measures.
6. To make soil surveys and related information more applicable and easily used in resource planning.
7. To strengthen the resource planning effort through more effective communication of the information provided by soil surveys and properly relating that information to modern technologies.

DEFINITION The soil potential index (SPI) is a numerical rating of relative suitability or quality (Soil Survey Staff, 1978; Vinar, 1980). It is derived from evaluations of soil performance, cost of corrective measures, and costs for continuing limitations. The soil potential index is expressed by the equation

$$SPI = P - (CM + CL)$$

where

P = index of performance or yield as a locally established standard

CM = index of costs of corrective measures to overcome or minimize the effects of soil limitations

CL = index of costs resulting from continuing limitations

Generally, the index number of the best soil map unit within a survey area is adjusted to 100 for the highest rating, and all other soils are expressed as lesser numbers or a percentage of the highest rating. In some cases index numbers can be greater than 100 if expressed in absolute units such as 120 bushels per acre (Kotlar, 1981).

EXAMPLES Soils with the highest potential index numbers are those with favorable attributes for the uses under consideration. The cost of corrective measures (CM) are the added costs above standard installation and management systems to overcome particular soil problems. The following list illustrates some soil limitations with general and specific measures to correct them:

Limitations	Broad Categories of Corrective Measures	More Specific Corrective Measures
Wetness	Drainage	Surface drainage
		Tile drainage
		Drainage land grading
Steep slope	Construction grading	Cuts and fills
Erodes easily	Erosion control	Permanent vegetation
		Grassed waterways
		Terraces
		Conservation tillage
High shrink-swell	Strengthened foundation	Reinforced slab
		Extended footings
		Moisture control
Floods	Flood control	Raised foundation
		Dikes
		Improved channels
Low strength	Supported foundation	Widened footings
		Extended footings
		Slab foundation
Droughty	Irrigation	Sprinkler irrigation
		Furrow irrigation
		Border irrigation

Continuing limitations (CL) are those that persist after corrective measures have been applied and have adverse effects on social, economic, or environmental values. Examples illustrating the derivation of continuing limitations (CL) include (Soil Survey Staff, 1978):

1. If the local performance standard is 2,000 pounds per acre, a potential production of only 1,500 pounds per acre from rangeland in a normal year, obtained through use of all feasible corrective measures for yield increase, is substandard by 500 pounds. Where P is 100, an appropriate index value for CL is

$$25\left(\frac{2000 - 1500}{2000} \times 100\right)$$

2. If flooding of a dwelling remains a probability after feasible measures are installed, an estimate of damage and inconvenience from a flood event divided by the frequency of flooding might provide an annual cost for conversion to index values. For example, damages of $6,000 might be estimated to result from floods occurring 1 year in 10. This represents an annual cost of $600 and a serious continuing limitation. An approximate value for CL might be 60 if the index for P is 100.

To calculate soil potential index numbers for dwellings without basements, for example (Soil Survey Staff, 1978), assume single-family residences each with 1,500 to 2,000 square feet of living space with slab construction, lawns, gardens, landscape design, play areas, on lot sizes of ¼ acre or less. Evaluating criteria are:

Factors Affecting Use	Degree of Limitation		
	Slight	Moderate	Severe
Depth to water table (inches)	> 30	18–30	< 18
Flooding	None	None	All years
Slope (percent)	0–8	8–15	> 15
Shrink–swell	Low	Moderate	High

Approximate index values and costs for correcting soil problems are:

Index Value	Cost Classes for Corrective Measures and Continuing Limitations
1	< 250
2	250–500
4	500–1,000
8	1,000–2,000
12	2,000–3,000
16	3,000–4,000
20	4,000–5,000

Corrective measures and their approximate costs for houses without basements are:

Corrective Measures	Costs	Index
Drainage of footing	$ 300–500	4
Drainage of footing and slab	600–800	7
Excavation and grading		
8–15% slopes	100–300	2
15–30% slopes	300–500	4
Rock excavation and disposal (fractured)		
0–8% slopes	1,000–1,400	12
8–15% slopes	700–900	8
Reinforced slab		
Moderate shrink–swell	1,500–2,000	17
High shrink–swell	3,600–4,200	39
Surface drainage (per lot)	100–200	2
Importing topsoil for lawn and garden	1,000–1,400	11

Each soil potential index number and each cost value, of course, must be formulated in specific or general terms most applicable to each survey area. Consultations among all people concerned are vital to the derivation of accurate and reliable data. These values presented here are examples only, and must be modified to fit each local situation. The numbers are most valuable when relatable to logical applications: these index values, for example, are one percent of estimated costs. The costs of continuing limitations are established for the 50-year life span of the dwelling to be compatible with costs of corrective measures.

When soil potential index numbers have been derived for all soil map units in a survey area for each specific use being considered, then general rating classes can be used to group soil potentials within the area. General definitions of such classes include:

1. *Very high potential.* Production or performance is at or above local standards, because soil conditions are exceptionally favorable. Installation and management costs are low and there are no soil limitations.
2. *High potential.* Production or performance is at or above the level of local standards. Costs of measures for overcoming soil limitations are judged locally to be favorable in relation to the expected performance or yields. Soil limitations continuing after corrective measures are installed do not detract appreciably

from environmental quality or economic returns.

3. *Medium potential.* Production or performance is somewhat below local standards, or costs of measures for overcoming soil limitations are high. Soil limitations continuing after corrective measures are installed detract from environmental quality or economic returns.

4. *Low potential.* Production or performance is significantly below local standards. Measures required to overcome soil limitations are very costly. Soil limitations continuing after corrective measures are installed detract appreciably from environmental quality or economic returns.

5. *Very low potential.* Production or performance is much below local standards. There are severe soil limitations for which economically feasible measures are unavailable. Soil limitations continuing after corrective measures are installed seriously detract from environmental quality or economic returns.

ASSIGNMENT Each student should derive soil potential index numbers for each soil map unit within the student's study area, for several land uses. Local contractors and other experts should be consulted to obtain cost figures and index numbers that are realistic and useful. Modifications should be freely made in the examples given here and in the references, to fit the local situations. Students should use their imagination and initiative freely, and make assumptions for feasible scenarios. A brief written report should be prepared explaining how the index numbers and cost figures were derived, and how the information can be used to evaluate the development potentials of different soil map units. A brief oral report (10 minutes) should also be presented by each student to the class. The instructor should arrange the assignment so that each student does not repeat the work of other students, and so that each student has ample opportunity to improvise in illustrating the application of soil potentials.

REFERENCES

Beatty, M. T., G. W. Petersen, and L. D. Swindale. 1979. Planning the uses and management of land. Monograph 21, American Society of Agronomy, Madison, WI. 1028 pages.

Kotlar, K. 1981. Application of the soil potential rating system to land use planning: A New York example. M.S. thesis, Dept. of Agronomy, Cornell University, Ithaca, NY. 77 pages.

Soil Survey Staff. 1978. Application of soil survey information: Policy guide. Parts I and II of Section 404 and Soil Potential Ratings of National Soils Handbook. U.S. Government Printing Office, Washington, DC. 45 pages.

Vinar, K. R. 1980. Soil potential ratings for the soils of Westchester County, New York. Soil Conservation Service, U.S. Dept. of Agriculture, White Plains, NY. 107 pages.

PURPOSE Measurements of soil variability are of tremendous help in understanding soils and the environment. Obviously, total significant variations within soil map units must be less than differences between soil map units—if the map delineations are to have practical applications. Soil variability is mostly a matter of detail of examination and soil map scale; as a general rule, soil investigations must be at a level of detail comparable to the level of detail to which the information is to be extended ("technology transfer") via soil map units. Thus, soils in agronomic experimental plots must be investigated in greater detail than in fields and farms, because the intensity of use and extension of data are greater. This exercise encourages the investigation of soil variability in order to more fully understand the nature of soil map units. Pages 41–52, 113–129, and 148–160 of the textbook and the references cited in this section should be studied in preparation for investigation of soil variability.

OBSERVATIONS While learning about field aspects involving uses of soils, people should make systematic observations about soil differences from place to place within short distances. In several 100-meter transects within a soil map unit, for example, make borings at 10-meter intervals to get a percentage estimate of soil variability and the "range in characteristics" within a soil map unit (see the exercise on "Soil Maps"). Commonly, in fields one observes vegetative responses due to soil differences such as those illustrated in Figures 86 and 87. Soil borings in the barren spots will reveal the reasons for the lack of corn growth, so that corrective measures may be applied. Such spots in fields are troublesome to farmers, because they cannot be treated in the same fashion as the rest of the field. Often tractors get stuck in the wet spots, and operations are delayed for the entire field until the small areas of wet soils have dried out. Most farmers have pride in the appearance of their fields and crops, and the soil variability is a source of discontent as well as an unproductive site.

In addition to transect borings, grid borings are extremely useful to characterize soil variability.

Select a grid pattern with a close enough spacing that a statistical characterization will be made of the soil variability. Thus, to evaluate soil variability across the barren spot shown in Figure 86, a grid spacing of several meters should be used. A 30-meter square area could be investigated, with borings every 3 meters—to determine the cause of the variations in corn growth. If experiment station plots are available for investigation, students and teachers should make several borings within each plot, and then relate the observations made of the soil borings to the long-term yields of the plots. Figure 88 shows an infrared photograph of experimental plots with great variability in soil drainage classes (and other soil properties) within each of the large plots. Figure 89 records borings to fragipan and mottles in smaller plots (12 ft X 40 ft). Obviously, soil differences are probably of greater significance to crop yields than any of the treatments in these experiments (see page 67 of the textbook). With careful study of the soils of each plot for each year, yields can then be related to specific soil properties. Fly and Romine (1964) have published a procedure for analyses of such experiments and the soil influences. Often, soil differences induce wider variations in yields than different treatments. Unfortunately, many agronomic experiments in the past have been laid out on the assumption that the soil was uniform, or else the experimental design was structured to mask the soil differences in the statistical analyses (see page 127 of the textbook).

Many reference works are available to assist students, teachers, and laypersons in statistical analyses of soil variability and the effects of soil differences. The first rule is to record carefully each soil observation and the precise location (Olson, 1975–1978) and to have systematic procedures and guidelines for programs to implement the data on soil variability (Olson, 1977). Beckett and Webster (1971) have published a review of many publications on soil variability. Webster (1977) has written an excellent comprehensive book on quantitative and numerical methods in soil classification and survey. Numerous authors (McCormack and Wilding, 1969; Pomerening and Knox, 1962; Powell and Springer, 1965) have investigated soil differences within map units; students should carefully study several of these papers and compare and contrast the research perspectives and the results. Some soil map units are highly variable, while others are more consistently mapped—the differences are due to soil and environmental differences and to many other factors. Linking of data to soil map units and understanding variability is the key to increased use of soil maps (Cipra et al., 1972; Westin, 1976). New methods of geostatistical analyses promise to increase our understanding of soil variability greatly in the future (Vieira et al., 1981; Yost et al., 1982).

As a class exercise, students should make borings in soils on a transect or grid basis, make careful notes on the soil observations and their locations, and subject the data to statistical analyses. The statistical analyses can range from simple calculation of means and ranges of values to geostatistical Kriging of closely spaced grid data. Student projects on soil variability will depend on the prior knowledge of statistics, and computer facilities available to the class. Although soils are variable as biological and natural entities, that diversity can be characterized and mapped and serves as the base for further understanding and use of soil surveys.

REFERENCES

Beckett, P. H. T. and R. Webster. 1971. Soil variability: A review. Soils and Fertilizers 34:1–15.

Cipra, J. E., O. W. Bidwell, D. A. Whitney, and A. M. Feyerherm. 1972. Variations with distance in selected fertility measurements of pedons of western Kansas Ustoll. Soil Science Society of America Proceedings 36:111–115.

Fly, C. L. and D. S. Romine. 1964. Distribution patterns of the Weld-Rago soil association in relation to research planning and interpretation. Soil Science Society of America Proceedings 28:125–130.

McCormack, D. E. and L. P. Wilding. 1969. Variation of soil properties within mapping units of soils with contrasting substrata in northwestern Ohio. Soil Science Society of America Proceedings 33:587–593.

Olson, G. W. 1975a. Fertility of soil map units in Cortland County, New York. Cornell Agronomy Mimeo 75–12, Cornell University, Ithaca, NY. 34 pages.

Olson, G. W. 1975b. Fertility of soil map units in Seneca County, New York. Cornell Agronomy Mimeo 75–13, Cornell University, Ithaca, NY. 20 pages.

Olson, G. W. 1976. Fertility of soil map units in Suffolk County, New York. Cornell Agronomy Mimeo 76–13, Cornell University, Ithaca, NY. 27 pages.

Olson, G. W. 1977. A proposal to formulate national procedures and guidelines for linking soil fertility evaluations to the soil survey in the United States. Cornell Agronomy Mimeo 77–34, Cornell University, Ithaca, NY. 19 pages.

Olson, G. W. 1978a. Fertility of soil map units in Franklin County, New York. Cornell Agronomy Mimeo 78–3, Cornell University, Ithaca, NY. 50 pages.

Olson, G. W. 1978b. Fertility of soil map units in Niagara County, New York. Cornell Agronomy Mimeo 78–4, Cornell University, Ithaca, NY. 35 pages.

Pomerening, J. A. and E. G. Knox. 1962. A test for natural soil groups within the Willamette catena population. Soil Science Society of America Proceedings 26:282–287.

Powell, J. C. and M. E. Springer. 1965. Composition and precision of classification of several mapping units of the Appling, Cecil, and Lloyd series in Walton County, Georgia. Soil Science Society of America Proceedings 29:454–458.

Vieira, S. R., D. R. Nielsen, and J. W. Biggar. 1981. Spatial variability of field-measured infiltration rate. Soil Science Society of America Journal 45:1040–1048.

Webster, R. 1977. Quantitative and numerical methods in soil classification and survey. Clarendon Press, Oxford, England. 269 pages.

Westin, F. C. 1976. Geography of soil test results. Soil Science Society of America Journal 40:890–895.

Yost, R. S., G. Uehara, and R. L. Fox. 1982. Geostatistical analysis of soil chemical properties of large land areas. II. Kriging. Soil Science Society of America Journal 46:1033–1037.

FIGURE 86/*Barren spot in cornfield in Tompkins County in central New York State. All areas within this field have the same climate and management conditions, so that the corn growth differences are strictly due to soil differences. Soil borings along transects or in a grid pattern would show that soils in the barren spot are more poorly drained, more clayey, more polluted, and have poorer soil structure and chemical composition in the surface. Subsoils in the lower barren spot are also poorer chemically and physically for root proliferation than in the better higher surrounding soils.*

FIGURE 87/*Soil differences in a cornfield in Lewis County in northern New York State. The entire cornfield has been uniformly planted and managed, but the soil differences are enormous, as illustrated by the differences in corn growth. Obviously, a soil boundary is located at the edge of the corn with a good stand. The part of the field in the foreground should be managed differently in another crop, or else the areas should be drained, limed, fertilized, and modified in other ways to improve the productivity. Costs of planting and growing the crop are the same for the area in the foreground as for the rest of the field, but the yields will be practically zero due to the poor soil conditions.*

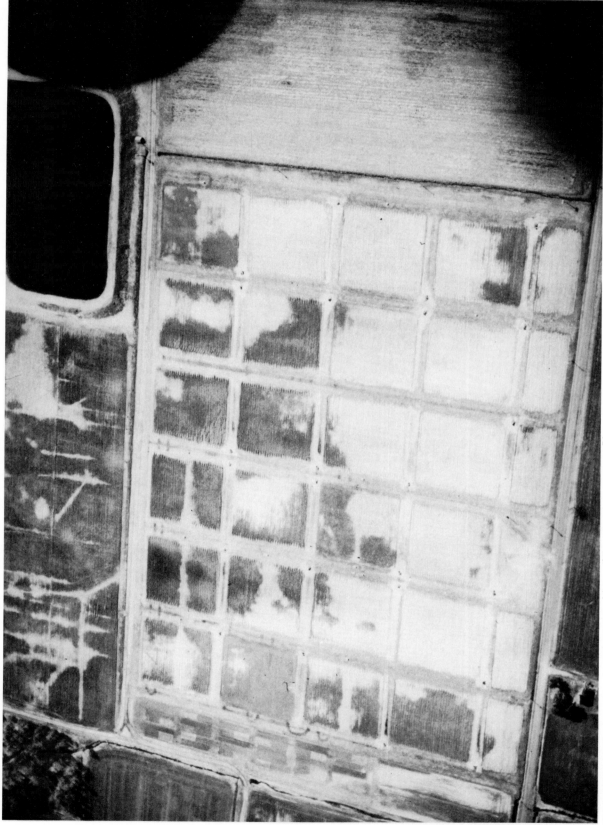

FIGURE 88/Oblique infrared aerial photograph of experimental plots near Aurora, New York. The dark areas within the plots are wet (poorly drained), and the light areas are drier (well drained). The black areas in the upper right portion of the photo are a pond and a part of the aircraft from which the picture was taken. (Photo by Stu Klausner.)

```
 12 12 10 18 21    18 18 28 14 15    10 12 14 20 11    13 14 14 17 23
 12 12 10 17 18    16 24 14 13 13    15 12  9 12 19    13 18 21 15 28
 14 11  9 20 19    17 30 16 13 13     8 11 10 15 17    19 17 20 15 25

 18 11 15 16 16    15 30 23 14 15    10 14 13 12 40    18 18 20 23 23
 18 11 16 14 19    16 30 15 11 18     8 16 16 29 21    26 25 20 16 27
 16 12 21 17 14    25 22 14 13 26     7 21 14 29 25    13 19 10 13 20

 16 12 12 12 14    13 15 36 36 14    12 10 16 11 12    17 16 21 17 29
 14 10  9 10 14    13 27 16 14 17    18 14 13 11 13    17 18 24 19 28
 20 11 14 23 14    16 17 15 11 18    14 17 32 11 10    15 22 18 17 19

 14 13  8 15 14    12 25 17 12 16    10 12 26 10 10    11 13 28 21 36
 12 15 14 12 14    13 32 11 19 10    11 13 20 15  9    11 12 24 27 34
 13 12 13 13 13    13 15 22 15 11    12 12 21 16 10    10 11 14 24 24

 18 12 13 14 22    14 24 10 14 14    17 17 28 11 12    21 12 13 13 19
 18 10 10 16 14    14 17 15 14 36    23 18 11 15  9    14 11 13 15 15
 19 11 11 10 17    12 22 31 12 16    14 29 13 12 18    17 30 14 14 12

 15 14 11 12 14    25 13 12 12 14    16 20 13 23 15    14 24 16 14 28
 17 14 17 15 19    12 17 12 26 14    19 27 13 30 21    16 15 12 17 28
 18 14 11 13 16    10 14 12 16 14    13 21 12 22 32    13 12 12 17 40

 18 14  9 12 16    12 16 28 24 16    16 20 10 11 20    18 20 16 24 36
 18 12 12 14 11    17 16 36 16 18    12 19 15 14 35    18 30 19 22 36
 18 16 10 33 16    15 12 13 15 16    15 19  9 13 28    12 30 16 19 36

 22 17 10 15 28    16 12 13 30 13    23 15 12 20 27    28 28 14 18 33
 18 15 12 12 14    23 12 12 33 15    19 22 23 30 23    30 22 32 18 36
 22 16 22 19 14    14 19 23 36 22    32 25 26 36 34    30 30 20 22 29
```

FIGURE 89/*Depth (in inches) to prominent mottles and fragipan at ends and middle of 12 ft × 40 ft experimental plots on Mt. Pleasant near Ithaca, New York. Shallow soils have been eroded, and deeper soils have alluvial and colluvial depositions.*

Sequential testing

PURPOSE Sequential testing is the process by which soils (and soil performances, yields, etc.) are sampled in relation to one another in accord with their natural positions in landscapes. Sequential testing offers great promise for future evaluations of soils and the environment because it enables soils to be isolated as the primary objects of study, but yet the relative landscape position is a principal element of consideration. The purpose of this exercise is to give students practice in achieving the sequential perspective of soils in landscapes by sampling soils and their performances in the sequences in which they occupy positions in the environment. In preparation for this exercise, students should study the previous exercises on Soil Maps, Yield Correlations, and Soil Variability; pages 127–129 of the textbook; and the references cited in this section.

Sequential testing is a term that has been used (Olson, 1981) to refer to sampling of soils across a landscape where the soils occupy "sequential" positions in relation to their respective location and gradient of properties in the topography of the landscape. Thus, soils in humid regions commonly have a "drainage sequence" in similar geologic materials (well drained, moderately well drained, somewhat poorly drained, poorly drained, very poorly drained); similar sequences can be observed over landscapes for soil texture, slope, pH, fertility, land use, crop growth, yields, and many other soil and land characteristics. Sequential testing of soils consists of studying the soils and their characteristics (including yields) according to the relationships each soil has with the others. Red pine (sensitive to soil wetness) will exhibit growth according to soil drainage classes, and sequential testing can measure the tree growth in relation to the soil depth to mottling and depth to seasonal water tables. Sequential crop sampling is particularly valuable because yield differences can be sampled which are strictly due to the soils (in single fields), where weather and management are constants (not variables in the sampling "experiment"). Thus, yield effects strictly due to soils can be isolated by sequential testing.

PROCEDURE Students should sample soils and yields within a "soil sequence" of their own choosing as illustrated in the idealized diagram of Figure 90. Soil borings should be made along transects to identify the soil sequences, and then yields should be measured at several selected places within each soil map unit (Figure 90). A "real" example of soils landscapes that would be well suited for sequential testing is given in Figure 91. Yield differences measured from different soils in the same field under the same crop would be strictly due to the sequence of soil differences; management and weather would be "constants" in this soil-yield sampling experiment. Similar samplings in a survey area will establish the effects of the soil differences in the behavior and performance of the soil map units.

OBSERVATIONS Many agronomic experiments in the past have been laid out on the assumption that the soil was uniform, or else the experimental design was structured to mask the soil differences in the statistical analyses. An excellent example of this approach has been provided by Fly and Romine (1964) for the USDA Central Great Plains Field Station at Akron, Colorado. The 25-acre plot area studied for 46 years of uniform management included 56 dryland rotations and tillage treatments on 152 plots, 8 X 2 rods in dimensions. The field appeared relatively uniform on the surface and was mapped as one soil type in a soil survey published in 1947. Mapping on a 100-foot grid at large scale on a base map with a 0.2-foot contour interval, however, showed that the area had soils that did indeed vary in color, thickness, texture, structure, depth to $CaCO_3$, and other significant properties. When the plots were indexed according to the eight soil map units finally identified, wheat production levels varied on those soils from 73 to 120. Wheat under continuous culture on deep soils produced 20 percent more paying crops and 55 percent more wheat above cost of production than did the thin soils in the experimental field plots of the same field. Soil differences induced wider variations under uniform treatment than did different treatments on a single soil map unit.

This study (Fly and Romine, 1964) is of extreme importance, because the concept of "technology transfer" from an experiment station to other areas is dependent on the soil identification. When the detail of delineation and characterization of soils is comparable to the detail of the research plot layout, only then can existing research fields and plot data be used to associate soil qualities with response to treatment. The reliability of application of research findings to other areas will depend on adequate characterization and evaluation of the soil in each plot. In the early days, when most experiment stations were initially sited, soil mapping was not sophisticated or detailed enough to give the kind of soil information now demanded by the needs for technology transfer to large areas characterized by modern detailed soil maps. Figures 88 and 89 provide additional examples of agronomic plot layout that disregarded the sequential testing of soils. Sequential testing "after the fact" (after the experiment has run for many years) is extremely difficult; future agronomic experiments should be designed strictly on the soil base considering extremely detailed mapping of the soil variability of the experimental areas.

Sequential testing is essentially a reversal of the idea of locating experiments or data collection points on uniform soils; sequential testing consists of deliberately locating experiments and data collection points on contrasting soils to evaluate selectively the effects of the soil differences. In sequential testing the soils are considered to be the basic resources, and the cropping and treatment design is fit into the soil characteristics of the landscape. Figure 90 illustrates schematically how sequential testing might work in practice. Experiments or data sampling points can be set up along a transect (or a grid) deliberately traversing soil boundaries. The experiment may be on a macro or micro scale to measure or sample land use, texture, drainage, slope, pH, fertility, or other characteristics of the soils. The soils isolated and studied, of course, depend on the nature of the landscapes and the soils that occupy their relative parts of landscapes. Along the sampling transect shown in Figure 90, complex experiments with monitoring points can be set up, or the sampling can be reduced to its simplest form by only "paired" or "triplicate" samplings of points within a single field. In Figure 91, place transect lines across soil boundaries in fields with the same crop. Soil and crop sequential samplings will show the correla-

tions between the soil characteristics and the plant growth and crop yields.

Examples of sequential testing can be found in the literature. Malo and Worcester (1974, 1975) and Malo et al. (1974) described soil and yield studies along a transect into a pothole in glacial till in the Red River Valley. A diagram of their transect and sampling points is given on page 128 of the textbook. One-foot contour intervals are marked on the map, and samplings were made of the soils and of barley and sunflower yields. The variable soils along the transect included Barnes (Summit Position-Udic Haploboroll), Buse (Shoulder Position-Udorthentic Haploboroll, Svea (Backslope Position-Cumulic Haploboroll), Hamerly (Footslope Position-Aeric Calciaquoll), and Tonka (Toeslope Position-Argiaquic Argialboll). Sunflower yields across the transect are plotted on page 129 of the textbook. Close relationships existed between the soil properties and the plant responses in the various landscape positions. These soils and related soils are repeated in sequences across large areas of these types of regional landscapes, so that the data of yield variabilities can be extended over large areas. In fields managed similarly, the different soils give very different responses and these responses are reflected in differences in quality and quantity of yields. Different sequential managements, of course, could produce a more uniform crop in these fields.

Sequential testing is also illustrated in a study of "slick spot" soils in Idaho (Lewis and White, 1964), where a trench 35 feet long was dug from a slick spot into the associated soils. Profiles were sampled at four points along the transect and showed clearly the sequence of depth of horizons, total salts, exchangeable sodium, clay mineralogy, and amorphous materials, which helped to explain the behavior and genesis of the soils in the slick spots and the surrounding landscape.

Many observations and a great deal of research can be related to soil sequences in landscapes and to opening up opportunities for sequential testing. A paper by Buol and Davey (1969) illustrates how tree growth is closely related to soil conditions, and how these sequences can be observed along cleared transects where tree heights can be readily seen. Many other situations, as where land-forming operations have cut and filled areas of different soils in fields (Phillips and Kamprath, 1973), offer possibilities to approach study of soils problems from the sequential testing perspective. Munn et al.

(1978) used soil sequences to explain soil effects on range habitat production in western Montana. Jenny (1980) reports on many soil sequence sampling situations.

The sequential testing approach to agronomic experiments has been used in plot work with equipment for application of fertilizer at a linear increase rate (Peck, 1976). Fox (1973) has described an experimental design where the fertilizer variable is increased in many small increments along corn rows. In this manner, individual plants can be handled as plots and the border effects are reduced. Replication can be obtained by repeating the layout and rotating it through 180 degrees. If a second factor is varied at right angles to the first, a well-defined response surface can be obtained.

The sequential testing approach can be expanded or contracted, as the landscape sequences stimulate ideas about experimental designs. Sopher and McCracken (1973) expanded sequential testing by sampling corn plots on a grid system with a sampling intensity of one to two plots per hectare. Locating plots on a grid system allowed the soils to be sampled as a continuum over the landscape. Sequential testing enabled continuous observations to be made rather than having the data clustered as semidiscrete observations around the means of a few selected soil series.

In contracting (simplifying) the sequential testing approach, numerous examples can be cited. Bouma and Hole (1971) used paired sampling of adjacent virgin and cultivated pedons to demonstrate soil changes over a century of cultivation. Binnie and Barber (1964) also used paired samplings to compare potassium release characteristics of alluvial and adjacent upland soils. Hart et al. (1969) studied good and poor stands of red pine separated by a fenceline, but used many monitored sampling points to evaluate the growth responses of trees in the "paired" soil situation. In Tennessee, considerable work has been done at experiment stations to measure yields on different soils of high contrast but close proximity (Fribourg et al., 1975; Fribourg and Reich, 1982; Table 63). In these sequential testings, effects of soil differences are clearly established through crop yields measured over long periods of time.

An excellent example of sequential testing in the ultimate sense is provided by Wallen and Jackson (1978). A plot measuring 197.5 X 43.9 meters was planted to 10,633 alfalfa plants (49 plants per row

spaced 0.9 meter apart, in 217 rows). The plants were then evaluated periodically for their survival during frost heaving over several seasons, on four soil map units which were delineated at large scale within the plant breeding plot. From an altitude of 105 meters, aerial photographs on 70-mm film were taken from a helicopter, to record the seasonal condition of each plant. Over several seasons the monitored plants clearly showed the effects of the soil conditions; well-drained soils supported a good population of plants, but poorly drained and moderately well-drained soils had high plant mortality. This method of detailed aerial surveillance even disclosed some inclusion areas within the well-drained soil map units which had restricted water movement, which in turn affected the alfalfa survival. This type of detailed sequential testing of soils is a very useful method that can provide keen insights into the behavior of plants due to soil sequence conditions; the study is an example of the use of initiative and inventiveness in the sequential testing of soils.

Many additional research papers show soil trends and variability across landscapes (Bornstein et al., 1965; Pomerening and Knox, 1962; Powell and Springer, 1965; Walker et al., 1968) which would be excellent subjects for sequential testing of soils and crop responses due to soils. As greater emphasis is placed on more detailed examinations of soils, sequential testing will become a vital part of the array of techniques used to fit the crops better to the soils through improved management. The sequential testing of soils will ultimately help to enable the improved "technology transfer" of data from experimental sites to large areas of similar soils that have been mapped in detail over the landscapes. Students, teachers, laypersons, and others working in soil survey interpretations will be hearing much more of sequential testing in the future.

REFERENCES

Binnie, R. R. and S. A. Barber. 1964. Contrasting release characteristics of potassium in alluvial and associated upland soils of Indiana. Soil Science Society of America Proceedings 28:387–390.

Bornstein, J., R. J. Bartlett, and M. Howard, Jr. 1965. Depth to fragipan in Cabot silt loam—variability in a characteristic area. Soil Science Society of America Proceedings 29:201–205.

Bouma, J. and F. D. Hole. 1971. Soil structure and hydraulic conductivity of adjacent virgin and cultivated pedons at two sites: A Typic Argiudoll (silt loam) and a Typic Eutrochrept (clay). Soil Science Society of America Proceedings 35:316-319.

Buol, S. W. and C. B. Davey. 1969. The soil, the tree. Journal of Soil and Water Conservation 24:149-150.

Fly, C. L. and D. S. Romine. 1964. Distribution patterns of the Weld-Rago soil association in relation to research planning and interpretation. Soil Science Society of America Proceedings 28:125-130.

Fox, R. L. 1973. Agronomic investigations using continuous function experimental designs—nitrogen fertilization of sweet corn. Agronomy Journal 65:454-456.

Fribourg, H. A. and V. R. Reich. 1982. Soils differ in yield potential. Crops and Soils 35:12-14.

Fribourg, H. A., W. E. Bryan, F. F. Bell, and G. J. Buntley. 1975. Performance of selected silage and summer annual grass crops as affected by soil type, planting date, and moisture regime. Agronomy Journal 67:643-647.

Hanna, W. E., E. B. Giddings, C. E. Rice, B. R. Laux, and J. E. Witty. 1981. Soil survey of Madison County, New York. Soil Conservation Service, U.S. Dept. of Agriculture, U.S. Government Printing Office, Washington, DC. 238 pages and 82 soil map sheets.

Hart, J. B. Jr., A. L. Leaf, and S. J. Stutzbach. 1969. Variation in potassium available to trees within an outwash soil. Soil Science Society of America Proceedings 33:950-954.

Jenny, H. 1980. The soil resource: Origin and behavior. Ecological Studies 37. Springer-Verlag, New York. 377 pages.

Lewis, G. C. and J. L. White. 1964. Chemical and mineralogical studies on slick spot soils in Idaho. Soil Science Society of America Proceedings 28:805-808.

Malo, D. D. and B. K. Worcester. 1974. Plant height and yield of sunflowers at different landscape positions. Farm Research 31:17-23. North Dakota Agricultural Experiment Station, Fargo.

Malo, D. D. and B. K. Worcester. 1975. Soil fertility and crop responses at selected landscape positions. Agronomy Journal 67:397-401.

Malo, D. D., B. K. Worcester, D. K. Cassel, and K. D. Matzdorf. 1974. Soil-landscape relationships in a closed drainage system. Soil Science Society of America Proceedings 38:813-818.

Munn, L. C., G. A. Nielsen, and W. F. Mueggler. 1978. Relationships of soils to mountain and foothill range habitat types and production in western Montana. Soil Science Society of America Journal 42:135-139.

Olson, G. W. 1981. A new concept in soil-plant-field research: Sequential testing. Soil Survey Horizons 22:5-8.

Peck, N. H. 1976. Equipment for application of fertilizer at a linear increase rate. Agronomy Journal 68:832-834.

Phillips, J. A. and E. J. Kamprath. 1973. Soil fertility problems associated with land forming in the Coastal Plain. Journal of Soil and Water Conservation 28:69-73.

Pomerening, J. A. and E. G. Knox. 1962. A test for natural soil groups within the Willamette catena population. Soil Science Society of America Proceedings 26:282-287.

Powell, J. C. and M. E. Springer. 1965. Composition and precision of classification of several mapping units of the Appling, Cecil, and Lloyd series in Walton County, Georgia. Soil Science Society of America Proceedings 29:454-458.

Sopher, C. D. and R. J. McCracken. 1973. Relationships between soil properties, management practices, and corn yields on South Atlantic Coastal Plain soils. Agronomy Journal 65:595-599.

Walker, P. H., G. F. Hall, and R. Protz. 1968. Soil trends and variability across selected landscapes in Iowa. Soil Science Society of America Proceedings 32:97-101.

Wallen, V. R. and H. R. Jackson. 1978. Alfalfa winter injury, survival, and vigor determined from aerial photographs. Agronomy Journal 70:922-924.

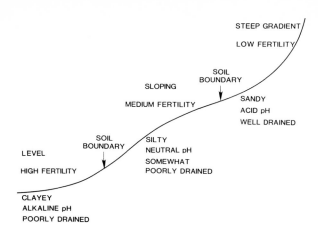

STEEP GRADIENT
LOW FERTILITY
SOIL BOUNDARY
SLOPING
MEDIUM FERTILITY
SANDY
ACID pH
WELL DRAINED
SOIL BOUNDARY
SILTY
NEUTRAL pH
SOMEWHAT POORLY DRAINED
LEVEL
HIGH FERTILITY
CLAYEY
ALKALINE pH
POORLY DRAINED

FIGURE 90/*Idealized diagram of sequences of soil characteristics for sequential testing. Soil slopes, fertility, texture, pH, drainage classes, land use, and yields are commonly correlated with landscape position in a sequence of properties and performances.*

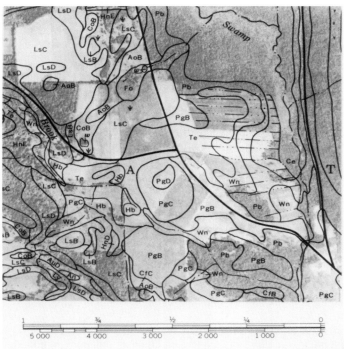

FIGURE 91/*Portion of soil map sheet 35 showing area north of Morrisville, New York, in Madison County (Hanna et al., 1981). A number of soil sequences in this area could be subjected to sequential testing. Adjacent soils under the same crop will have different yields due to soil sequence differences. Yield measurements would quantify the effects of the sequential soil properties. Some of the soils in the area include: PgB Palmyra, Te Teel, Wn Wayland, Pb Palms, Ce Carlisle, LsC Lansing, AoB Appleton, and Hb Hamlin.*

TABLE 63/*Yields (metric tons of dry matter/hectare) of paired samplings of corn and sorghum from early plantings in Tennessee in 1969 (adapted from Fribourg et al., 1975).*

Location	Soil	Corn	Sorghum
Knoxville	Sequoia	9.03	19.44
	Statler	18.16	25.64
Springfield	Dickson	15.63	19.95
	Huntington	28.22	35.03
Spring Hill	Maury	9.82	14.68
	Emory	15.33	25.96
Martin	Grenada	13.83	29.85
	Collins	15.99	32.80
Jackson	Dexter	7.38	19.05
	Loring	8.46	17.66
Crossville	Tilsit	17.01	13.47
	Hartsells	14.01	15.47

Land uses and soils

PURPOSE Statistical relationships between land uses and soils must be investigated much more in the future than they have been in the past, to improve uses of soil surveys. Most current soil survey interpretations are based on causal relationships that soil surveyors have observed while mapping soils, but few statistics are available to quantify the observations. This exercise outlines techniques of relating land uses to soils via statistics and computers, and gives examples of such studies. People making these correlations (students, teachers, laypersons) should study pages 98–104 and 114–118 of the textbook. Procedures are outlined, and examples are given from several studies done by students for pilot projects in New York State. Each user of this Field Guide should acquire some experience in making correlations between land uses and soils. Land uses in most places are a reflection of trial-and-error methods of land development of the past, and by statistical comparisons of land uses and soils we can see how corrections should be made to improve land uses and soil uses in the future.

PROCEDURE Study several books about land uses and aerial photographic interpretations (Bartelli et al., 1966; Beatty et al., 1979; Davidson, 1980; Johannsen and Sanders, 1982; Petersen and Beatty, 1981) to achieve an understanding of the present perspectives of land uses and soils. Soil survey interpretations are based on the field properties of the soil map units, and land use is mapped mostly by remote-sensing techniques with computer, cartographic, and photographic enhancements.

Select an area of several square miles with highly contrasting soils and land uses. Adjust both the soil map and the land-use map for the area to the same scale (this can be done photographically or cartographically). Then make a transparent grid overlay marked off in squares equivalent to 1 acre, 1 hectare, or other area convenient to work with at the selected scale. For example, a grid of about 1/8-inch squares at a scale of 1:20,000 is approximately a grid of acre cells; a grid of 6.2-mm squares at a scale of 1:15,840 is approximately a grid of hectare cells; and so on. The grid cell size depends on the detail of the soil and land-use patterns on

the maps, and should always be smaller than the minimum-size delineations so that no map unit is lost. Put a dot in the center of each square. Put the transparent grid overlay over the soil map and then over the land-use map, recording the soil and land use under each dot in each cell. Be sure that each overlay cell recording is in the same cartographic position on the maps, and that the transparent overlay is not shifted as the counting progresses. A summary of the soil and land-use relationships helps to explain soil effects on land-use patterns, and suggests past errors and future corrections.

LAND-USE MAPS Land-use maps are generally made from aerial photographic interpretation with relatively little fieldwork; soil maps, in contrast, are made mostly from fieldwork by digging in the soils. Land uses are often readily apparent from surface and aerial viewing, but soils may be different beneath the same land use or have only subtle (but important!) differences. Many schemes and legends have been developed to map land uses (Johannsen and Sanders, 1982; Petersen and Beatty, 1981). In New York State, land uses have been mapped on aerial photographs at a scale of 1:24,000 and uses of the maps are administered as part of the program of the Resource Information Laboratory at Cornell University. Part of the legend of the land use mapping for New York State follows:

Ac	Agricultural cropland
Ai	Inactive agricultural land
Ap	Pasture land
At	High-intensity cropland
Ao	Orchards
Fb	Brushy forest
Fn	Natural forest
Fp	Plantation forest
Or	Outdoor recreation
Wb	Brushy wetland
Ww	Wooded wetland
Cs	Commercial use
P	Public use
Eg	Sand and gravel pits
Ta	Airport
Rc	Farm labor camp
Ih	Heavy manufacturing

PONY HOLLOW As a student project in Agronomy 497 (Special Topics in Soil Science) at Cornell University, S. M. Frappier selected the Pony Hollow area southwest of Ithaca shown on page 103 of the textbook for a computer study of relationships between land uses and soils. A summary of her project report is presented here as an example and as a guide to other students doing similar land-use and soils correlations. Figure 92 is the overrun soil map (without air photo background) of the area. A copy of the land-use map (Figure 93) was redrafted from a scale of 1:24,000 to a scale of 1:20,000 (the same scale as the soils map). The data were collected by placing the transparent 1/8-inch grid (approximately 1 acre) over the soils map (Figure 92) and then over the land-use map (Figure 93). At each dot in the center of each grid square, the soil map unit symbol and the land-use symbol for the area under the dot were recorded on file cards and arranged alphabetically as to soil and land use. A matrix of the numbers of observations was compiled for soils in rows and land uses in columns (Table 64).

After the data were collected (Table 64), the soil map units that were recorded were classified according to the slope, depth to seasonal high water table, permeability, texture, and presence or absence of a fragipan. These characteristics were recorded in another matrix with the characteristics as columns and the soil map units as the rows. When compared with the land uses, this tabulation enabled correlations not only between land uses and soils, but also between land uses and specific soil properties.

The final step was writing a computer program to rearrange the information in the matrices. The two input matrices were the grid point data and the soil characteristics. The output matrices were six in number: one matrix of the percent of each land use that is on each soil, and five other matrices, one for each soil characteristic, that shows what percent of the land for each use is made up of each characteristic classification. Many different computer programs can handle these types of matrix correlations.

Figures 94 to 98 statistically summarize the land uses and soils relationships in the Pony Hollow area. Remember that these data apply only to the soils and land uses in the Pony Hollow area, but similar studies could be done in other geomorphic areas or for the state or nation as a whole. Agricultural cropland (Ac, Figure 94) mostly occupies the gentler slopes; more than 75 percent of the cropland has slopes less than 10 percent. However, only about 20 percent of the inactive agricultural land (Ai) has slopes less than 10 percent; about half (50 percent) of the abandoned farmland (Ai) has slopes above 15 percent. About 86 percent of the commercial land (Cs) is located on soils with slopes less than 15 percent, but some commercial use (15%) occupies the steepest slopes (35 to 70 percent). Forest uses (Fb, Fn, Fp) and outdoor recreation (Or) occupy mostly the steeper soil slopes, with 42 percent of the outdoor recreation use occupying 35 to 70 percent slopes. Wetlands (Wb, Ww) are mostly nearly level, except for wet sloping soils and seep spots; the brushy wetland (Wb) consists of soils that were once cleared and farmed or pastured, but are now abandoned for farming. Clearly, the best soil slopes are most intensively used, and a good proportion of the steeper soils once farmed have now been abandoned. The properties of the soils help to explain the land-use shifts that have occurred. Probably much of the original land clearing was done without very good knowledge of the soil conditions.

The best soils for most uses in Pony Hollow have gravelly or silt loam textures (Figure 95), and are associated with good permeability and drainage. Channery silt loam soils are associated with the acid upland glacial till, and 56 percent of the inactive agricultural land (Ai) is on such soils. Obviously, the acid channery silt loam soils with a dense fragipan in the subsoils were an inadequate resource base for the farmers who settled in these places. One can easily speculate on the human tragedies resulting from the inadequate soils knowledge that came to the farm families before they abandoned these lands. In balance, almost 70 percent of the natural forest (Fn) occupies the channery silt loam soils that settlers avoided; almost 70 percent of the brushy forest (Fb) represents forest regrowth on these soils after farms were abandoned. The soil properties provide a statistical record of the reasons for abandonment, and the historical and archaeological records support the conclusions based on the soil characteristics. One finds many abandoned stone fences and farmsteads in the forest regrowth when hiking through this area. Evidences of soil mismanagement also abound in eroded fields and sediment accumulations in old fencerows. The brushy wetland (Wb) has alluvial accumulations from the eroded uplands. Only 9 percent of the currently active cropland (Ac) is located on the channery silt loam soils.

Most (60 percent) of the agricultural cropland (Ac, Figure 96) in Pony Hollow has soils with moderate to rapid permeability. Most commercial use (Cs, 85 percent) also occupies the moderately permeable soils. The wetlands (Wb, Ww) are so located because of high water tables not strictly related to permeability (slow in glacial till, rapid in glacial outwash). Subsoils in glacial till in the uplands have slow permeability in the dense fragipan, and often water flows laterally downslope above the pan and comes to the soil surface downslope in seepage spots. Soils in gravelly glacial outwash in the valley are mostly rapidly permeable and well drained.

Depth to high water table (Figure 97) has a major influence on land use in Pony Hollow. Agricultural (Ac, Ap) and commercial (Cs) uses are best where the water table is below 15 inches. Far more of the pasture land (Ap) has a high water table above 5 inches — 26 percent versus 6 percent for cropland (Ac). Soils good for farming, but often too wet to plow, can be used for pasture. Wet soils can be improved through drainage. More than 70 percent of the brushy wetland (Wb) has soils with high water tables near the surface. Some better drained soils in parts of fields may have been abandoned when wet fields (Wb) became no longer profitable to plow or pasture.

Fragipans are common in soils in acid glacial till in the uplands around Pony Hollow. Figure 98 summarizes the effects of the dense impermeable acid fragipan on land use. Currently, active agricultural cropland (Ac) is mostly located on the soils in gravelly glacial outwash without fragipans. About 40 percent of the inactive agricultural land (Ai) has a fragipan, but has been abandoned for farming. Even pastureland (Ap) is not common on soils with fragipans. Most soils with fragipans are occupied by forest (Fb, Fn, Fp), are used for outdoor recreation (Or), or are classified as wetlands (Wb, Ww). Without a doubt, the soils of the Pony Hollow area have had and will have a dominant influence on the past, present, and future land uses.

MELVIN BROOK As another student project to correlate land uses and soils, J. D. Sullivan selected the Melvin Brook drainage basin of about 2,000 hectares in Wayne County near Clyde, New York,

where soils are mostly on nearly level slopes. He reduced the soil map at 1:15,840 scale to a scale of 1:24,000 to have it on the same scale as the topographic maps of the U.S. Geological Survey. The land-use map was also at a scale of 1:24,000. He used a grid cell size equivalent to 1 hectare, but did not use a computer in his study. Soil properties considered were slope, permeability, texture, soil reaction, drainage class, depth to the seasonal high water table, trafficability, and total available water capacity. Agriculture is the principal land use in the drainage basin, with cash crops such as corn, snap beans, potatoes, strawberries, and wheat in addition to some support crops for the dairy industry. The watershed also contains about 300 hectares of wetlands associated with the stream channel, and a number of scattered woodlots. Many of the soils are wet and have been artificially drained.

A transparent grid overlay of 1-hectare cells was laid on the soil map and land-use map with each row and column of the grid labeled so that exact locations within the watershed could be specified. The soil and land use beneath each dot in the center of each cell were recorded and cross-classified by soil map unit and land-use symbol. Color maps were drawn for each of the soil properties considered. Table 65 provides a matrix of the soil and land-use data recorded for the cells.

More than 800 hectares in the watershed are Canandaigua soils (Ca). Other dominant soils are Ontario (OnB, C, D), Hilton (HnA, B), Minoa (Mn), and Rhinebeck (RaA). The most common land uses are cropland and pasture (Ac, Ap) occupying 53.4 percent of the total area. Inactive agricultural land (Ai), high-intensity cropland (At), brushy forest (Fb), and wooded wetland (Ww) are common (Table 65). Trafficability, drainage class, available water capacity, slope, pH, permeability, and texture are closely related to land use.

Trafficability of soils in the Melvin Brook watershed are mostly good or moderate under agricultural land uses; forests and wetlands, in contrast, commonly have soils with moderate to poor trafficability. Miscellaneous land uses (Cs, P, Eg, Ta, Rc, Ih) commonly have soils with good trafficability, rapid permeability, and coarser textures. Drainage and available water in many of the soils are poor, but management is good and many of the wet soils have been improved with artificial drainage. Over half of the orchard land in the watershed has moderate to well-drained soils.

High-intensity cropland is mostly on poorly drained soils, but tile-drained. Wetlands are located on the finer-textured soils. Only 142 hectares of soils is strongly acid; most soils are slightly acid or neutral; inactive agricultural land, brushy forest, and orchards are more common on the more acid soils.

The Melvin Brook watershed land uses and soils correlations contrast well with the Pony Hollow correlations in different geomorphic regions. Land uses are more closely correlated with soils in the Pony Hollow area because the soil map units are more highly contrasting. In the Melvin Brook watershed the soils have fewer slope contrasts and more widespread wetness problems that have been somewhat overcome by artificial soil drainage. Nevertheless, the soil factors are important in determining land uses and management in both areas, and help greatly to explain the history of land use and future potentials for improvements. With greater knowledge of land uses and soils correlations, planning and development can be improved considerably in the future.

CANADARAGO COMPUTER STUDY Pages 114–118 of the textbook describe a different computer study where soils, land use, and topography in a drainage basin were modeled in relation to the soil-loss equation (see pages 108–113 of the textbook). In this study 10-acre cells were used in the 41,930-acre watershed. Soil loss, deposition, and movement were determined for each cell, and from cell to cell, to streams and the lake. Then phosphorus attached to the soil particles was determined and modeled in relation to eutrophication of the lake. This study is much more complex than the more simple land uses and soils correlations, because it involves dynamic processes and computer modeling. The Canadarago computer study consisted of work for the M.S. and Ph.D. degrees by G. F. Kling (Kling and Olson, 1975).

RECOMMENDATIONS These land uses and soils correlations illustrate only a few of the possibilities for making resource inventories and correlations to soils maps. In a similar fashion, geology maps, groundwater maps, vegetation maps, wildlife maps, demographic maps, meteorological maps, tax maps, pollution maps, engineering maps, watershed maps, economic maps, planning maps, and many others can be correlated to the soil conditions. Scale of maps and size of cells can be modified to fit the inventory and resource problems and potentials; page 102 of the textbook illustrates land uses (agricultural prosperity) and soils correlations made using grid cells of 1 square mile on maps at a scale of 1:250,000. As computer technology improves, it will become increasingly easier and cheaper to make correlations of soils to other cartographic factors expressed on maps. Everyone interested in soils and the environment should become familiar with the techniques and possibilities outlined in this exercise and these examples.

REFERENCES

Bartelli, L. J., A. A. Klingebiel, J. V. Baird, and M. R. Heddleson (Editors). 1966. Soil surveys and land use planning. American Society of Agronomy, Madison, WI. 196 pages.

Beatty, M. T., G. W. Petersen, and L. D. Swindale (Editors). 1979. Planning the uses and management of land. Monograph 21, American Society of Agronomy, Madison, WI. 1028 pages.

Davidson, D. A. 1980. Soils and land use planning. Longman, New York. 129 pages.

Higgins, B. A. and J. A. Neeley. 1978. Soil survey of Wayne County, New York. Soil Conservation Service, U.S. Dept. of Agriculture, U.S. Government Printing Office, Washington, DC. 210 pages and 71 soil map sheets.

Johannsen, C. J. and J. L. Sanders. 1982. Remote sensing for resource management. Soil Conservation Society of America, Ankeny, IA. 665 pages.

Kling, G. F. and G. W. Olson. 1975. Role of computers in land use planning. Information Bulletin 88, New York State College of Agriculture and Life Sciences, Cornell University, Ithaca, NY. 12 pages.

Petersen, G. W. and M. T. Beatty (Editors). 1981. Planning future land uses. Special Publication 42, American Society of Agronomy, Madison, WI. 71 pages.

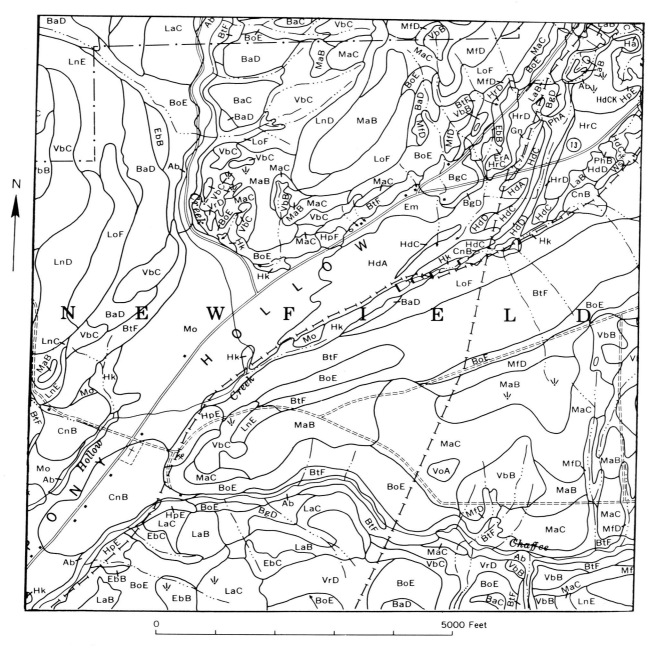

FIGURE 92/*Detailed soil map of Pony Hollow area southwest of Ithaca, New York, without air photo background (see page 103 of the textbook). The legend for this soil map is presented in Table 6.*

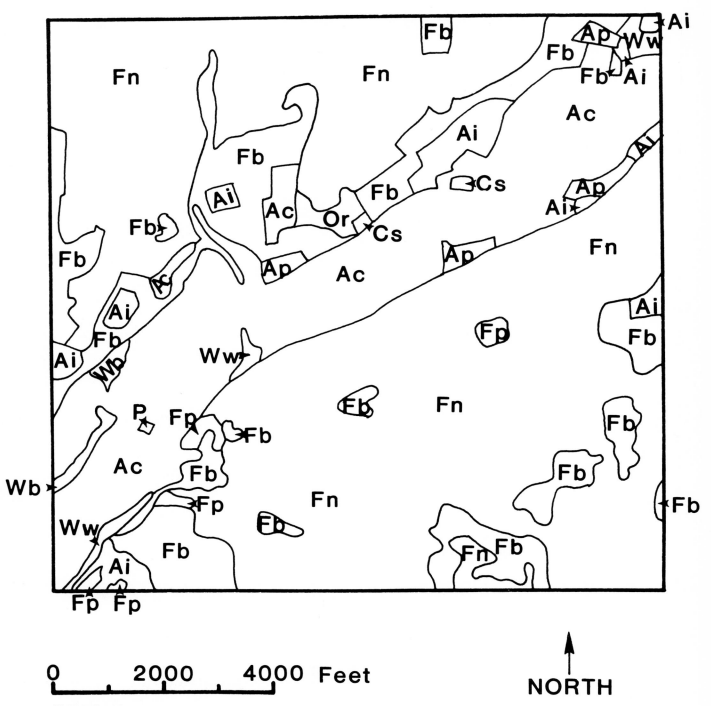

FIGURE 93/*Land-use map of Pony Hollow area southwest of Ithaca. Using a grid overlay, this map was statistically compared with the detailed soils map shown in Figure 92.*

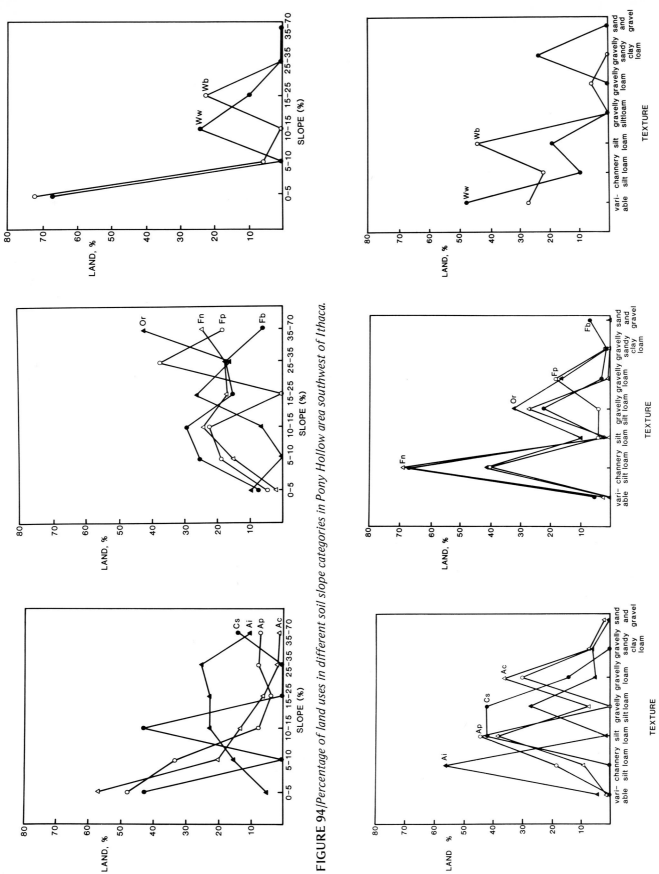

FIGURE 94/Percentage of land uses in different soil slope categories in Pony Hollow area southwest of Ithaca.

FIGURE 95/Percentage of land uses with different soil textures in Pony Hollow area southwest of Ithaca.

170

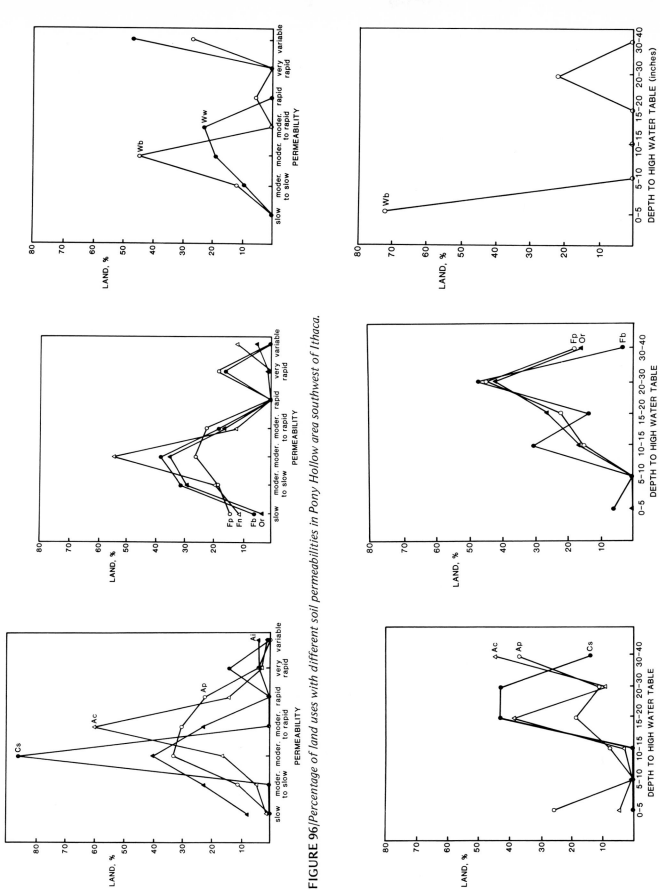

FIGURE 96/Percentage of land uses with different soil permeabilities in Pony Hollow area southwest of Ithaca.

FIGURE 97/Percentage of land uses showing depth to high water tables in soils of each in Pony Hollow area southwest of Ithaca.

171

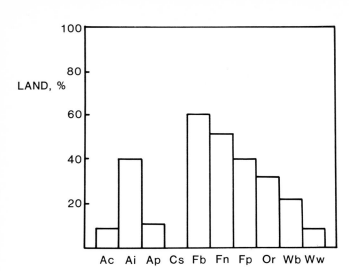

FIGURE 98/*Percentage of land uses that have a fragipan in soils of each in Pony Hollow area southwest of Ithaca.*

TABLE 64/*Matrix table of acres of each land use for each soil in Pony Hollow area southwest of Ithaca (see Table 6 for legend for soil map units).*

Soil	Land Use										Total
	Ac	Ai	Ap	Cs	Fb	Fn	Fp	Or	Wb	Ww	
Ab	5	0	0	0	21	33	0	0	5	8	72
BaC	4	0	0	0	0	20	0	0	0	0	24
BaD	1	6	0	0	7	79	0	1	0	0	94
BgC	3	6	0	3	0	0	0	0	0	0	12
BgD	10	0	0	0	5	13	0	0	0	0	28
BoE	1	12	0	0	63	175	5	5	0	0	261
BtF	6	4	0	0	23	196	5	5	0	0	239
CnB	78	0	6	0	1	1	0	0	1	0	87
EbB	3	8	0	0	9	20	2	0	0	0	42
EbC	0	0	0	0	7	7	0	0	0	0	14
Em	9	1	0	3	0	0	0	3	0	0	16
EcA	3	0	0	0	0	0	0	0	0	2	5
Fm	1	3	0	0	0	0	0	0	0	0	4
Gn	5	0	0	0	0	0	0	0	0	0	5
HdA	80	0	1	0	0	0	0	0	0	0	81
HdC	25	3	2	0	3	0	0	0	0	5	38
HdCK	13	2	0	0	2	0	0	0	0	0	17
HdD	10	0	0	0	0	0	0	0	0	0	10
Hk	19	0	4	0	4	4	0	0	8	4	43
HpE	8	3	1	0	4	6	5	0	0	0	27
HpF	0	0	0	1	0	0	0	5	0	0	6
HrC	23	0	0	0	2	0	0	0	0	0	25
HrD	11	1	0	0	3	0	0	0	0	0	15
LaB	4	4	3	0	10	27	3	0	0	0	51
LaC	0	0	0	0	8	56	4	0	0	0	68
LnC	0	4	0	0	1	0	0	0	0	0	5
LnD	0	0	0	0	24	44	0	0	0	0	68
LnE	0	5	0	0	3	61	0	0	0	0	69
LoF	1	4	2	0	3	148	0	3	0	0	161
MaB	21	0	0	0	53	124	0	0	0	0	198
MaC	2	0	0	0	35	162	0	0	0	0	199
MfD	0	11	0	0	10	59	0	7	4	2	93
Mo	181	0	5	0	4	0	1	0	0	0	191
PhA	15	0	0	0	0	0	0	0	0	0	15
PhB	3	0	0	0	0	0	0	0	0	0	3
VbB	3	0	0	0	30	42	0	0	0	0	75
VbC	5	3	0	0	60	91	2	2	0	0	163
VoA	1	0	0	0	0	3	0	0	0	0	3
VrD	4	0	0	0	12	36	0	0	0	0	52
Ws	0	0	3	0	0	0	0	0	0	0	3
Total	557	80	27	7	407	1407	27	31	18	21	2582

TABLE 65/Matrix table of hectares of each land use for each soil in Melvin Brook drainage basin in Wayne County near Clyde, New York (see Higgins and Neeley, 1978, for legend of soil map units).

Soil	Land Use																Total
	Ac	At	Ap	Ai	Ao	Fb	Fn	Fp	Ww	Wb	P	Eg	Ta	Cs	Rc	Ih	
Ca	392	127	25	27	2	43	54	1	130	28	0	0	0	0	0	0	829
HnA	4	0	0	0	0	0	0	0	0	0	0	0	0	0	0	0	4
HnB	80	11	3	25	1	14	4	0	6	1	2	0	0	0	0	0	147
OnB	95	20	9	29	6	15	5	0	4	2	2	0	0	0	0	0	187
OnC	20	4	0	7	0	5	2	0	0	0	2	0	0	0	0	0	40
OnD	5	0	1	2	0	5	0	1	0	0	0	0	0	0	0	0	14
OSE	3	0	0	0	0	5	0	0	0	0	0	1	0	0	0	0	9
PcA	30	0	0	0	0	0	0	0	0	0	0	1	2	0	0	0	33
PcB	42	6	1	2	0	4	0	0	0	0	3	0	1	0	0	0	59
PcC	3	0	0	0	0	0	0	0	0	0	0	0	0	0	0	0	3
AgA	4	0	0	0	0	0	0	0	0	0	0	0	0	0	0	0	4
AgB	25	0	2	2	1	11	0	0	0	0	0	1	0	0	0	0	41
AgC	7	0	0	0	0	0	0	0	0	0	0	0	0	0	0	0	7
MdB	0	0	0	0	0	2	0	0	0	0	0	0	0	0	0	0	2
MdC	2	0	1	0	0	0	0	0	0	0	0	0	0	0	0	0	3
Cd	0	0	0	0	0	2	0	0	14	6	0	0	0	0	0	0	22
CrB	1	0	0	0	0	0	0	0	3	1	0	0	0	0	0	0	5
PoA	10	0	0	4	0	0	0	0	1	0	0	0	0	0	0	0	15
PoB	2	0	0	0	0	0	0	0	0	0	0	0	0	0	0	0	2
WnA	13	0	0	0	0	0	0	0	0	0	0	0	0	0	0	0	13
WnB	4	0	0	0	0	0	0	0	0	0	0	0	0	0	0	0	4
OaB	2	0	0	0	0	6	0	0	0	0	0	0	0	0	0	0	8
BoB	0	0	0	0	0	0	0	0	0	0	0	0	0	0	1	0	1
Mn	136	0	32	32	0	14	8	0	12	0	0	0	0	2	1	0	237
Ap	2	0	0	0	0	0	0	0	2	0	0	0	0	0	0	0	4
Fr	2	0	0	0	3	1	0	0	0	0	0	0	0	0	0	0	6
RaA	98	0	7	7	0	11	14	0	26	0	0	0	0	0	0	0	163
ElA	17	0	1	1	0	2	1	0	1	2	0	0	0	1	0	2	28
ElB	20	0	0	0	0	0	0	0	0	0	0	0	0	0	0	0	20
Lm	0	0	3	3	0	1	0	0	8	0	0	0	0	0	0	0	15
Ma	1	0	2	2	0	0	1	0	27	0	0	0	0	0	0	0	33
Wa	63	0	5	5	0	3	7	0	0	0	0	0	0	0	0	0	83
OvB	1	0	0	0	0	2	0	0	0	0	0	0	0	0	0	0	3
CoA	0	0	1	1	0	0	0	0	0	0	0	0	0	1	0	0	3
CoB	0	0	1	1	0	3	0	0	0	0	0	0	0	0	0	0	5
Pa	0	0	0	0	0	5	0	0	1	0	0	0	0	1	0	0	7
C.F.L.*	2	0	2	0	0	5	4	0	0	0	0	0	0	2	0	0	15
FW	0	0	0	0	0	0	0	0	0	3	0	1	0	0	0	0	4
Gravel pit	0	0	0	0	0	0	1	0	0	0	0	0	0	0	0	0	1
Total	1086	168	97	150	13	159	101	2	235	43	9	4	3	7	2	2	2081

*Cut and fill land.

Tragedy of the commons

PURPOSE This exercise is philosophical, to make students, teachers, and others think about the full implications of soil and environmental use and abuse. The concept of the "tragedy of the commons" is of utmost importance in understanding the "carrying capacity" of soils and the environment. People are encouraged to study the examples and references given, and to search for illustrations of the "tragedy of the commons" in their own locality and project areas. Textbook chapters on erosion control (pages 105-118), archaeological considerations (pages 130-142), and definitions on pages 157 and 167 of the textbook should be studied in preparation for this exercise.

DEFINITION The concept of the "tragedy of the commons" was discussed in a thoughtful article by Hardin (1968). The article has received much academic acclaim, and an educational movie has even been made to explain the subject. Many authors and editors (Baden, 1980; Coates, 1972; Crowe, 1969; Horsfall, 1972; Love and Love, 1970; Olson, 1981; Roos, 1971) have expanded on the examples and principles of the concept. Basically, the concept (Hardin, 1968) is that human beings are incapable of managing their resources when greed, population, and exploitation are uncontrolled. The "commons" is a community pasture, park, forest, soil area, or other resource that can be used by all. The "commons" system works well when the community population is small and the resource supply is large, but the system fails when productivity and environmental quality decline as the population expands and exploits the commons excessively. The tragedy "resides in the solemnity of the remorseless working of things" as the environmental system reverts to a low level or collapses as a result of over-exploitation. Most human activities and problems (e.g., nuclear arms race, pollution, slums, wars, greed) can be related to the concept of the "tragedy of the commons."

In New England villages (Horsfall, 1972), each community had a "commons" (e.g., Boston Commons):

> In those days, of course, almost eveyone had to have a cow because there were no dairies; hence, the villages created the commons as a place to pasture cows. Everyone had a right to pasture a cow on the commons. This system worked fine as long as the village was small and the commons was big enough to support the cattle. The village population grew and the number of cattle increased until eventually there was insufficient space for all the cattle to graze....
>
> Sooner or later a drought struck and put the commons under stress. The cattle would give increasingly smaller amounts of milk and the children who depended on it for their sustenance would not have quite enough to drink. Eventually, under this plan, the commons would collapse.
>
> The fundamental characteristic of this parable is the collision of two freedoms. Everyone had a right to put a cow on the commons, and everyone's children had a right to a supply of milk. The villagers had an infinite freedom, but they had only a finite commons. And this is where the tragedy of the commons began....

Baden (1980) further elaborated on the tragedy of the commons:

> Several herdsmen graze their animals on the commons free of charge and each herdsman may put as many animals on the pasture as he sees fit. So long as the total number of animals is below the carrying capacity of the commons, a herdsman can add an animal to his herd without affecting the amount of grazing for any animals including his own. But beyond this point, the addition of another animal has negative consequences for all users of the common pasture including the herdsman. A rational herdsman, seeing that the benefit obtained from the additional animal accrues entirely to himself but that the effects of overgrazing are shared by all the herdsmen, adds an animal to the commons. For the same reason, he decides to continue adding animals, as do the other herdsmen. Each, through an individual calculus, realizes that if he refrains from adding the additional animals, he will have to absorb a greater share of the external costs generated by the maximizing herdsmen. He becomes a "sucker" while others obtain an essentially free ride at his expense. The process of adding

animals may continue until the ability of the commons to support livestock collapses entirely. Empirical evidence such as the West African Sahel stand in awesome testimony of the validity of the "logic of the commons." It should be noted that an individual herdsman acting alone cannot save the commons through his own actions since he is not alone in endangering it. Unilateral restraint only assures a herdsman a smaller herd, not a stable pasture. The "tragedy of the commons" is not simply the fact that the commons is destroyed, but that rational individual action may produce consequences that leave everyone worse off.

Overgrazing problems are greatest in societies where wealth is determined by numbers of animals regardless of condition, and cost of feed by grazing on the "commons" is zero. The Sahel region in Sub-Sahara Africa is one example (Anonymous, 1974; Cowell, 1982; Franke and Chasin, 1980; Wade, 1974). The area is marginal for grazing especially in drought periods, and the results of the "tragedy of the commons" are disastrous to both animals and human beings.

MOUNTAINS Forested mountains and hills provide another example of the "tragedy of the commons" (Coates, 1977; Eckholm, 1975, 1976; Gentry and Lopez-Parodi, 1980; Lowdermilk, 1978; Riding, 1974; Thomas, 1980; Thomas, 1965; Webster, 1982). In a natural state, forests protect steep slopes from erosion and assist in gradual and sustained groundwater recharge. As populations increase, however, trees are cut for firewood and construction of houses and other buildings. Roads, grazing, and farming on slopes increases erosion and runoff. Finally, the productivity collapses and floods and sedimentation deposits originating in the highlands damage the lowlands also.

In 1982, the first international multidisciplinary meeting on the subject of mountain resources (Webster, 1982) summarized the process of the "tragedy of the commons" in the deterioration of soils and the environment in mountains:

The underlying cause of such degradation of alpine ecosystems is often a "domino" scenario that begins with overpopulation, usually the result of human migration from crowded lowland cities. This leads to deforestation to satisfy growing timber and fuel needs, then to the expansion of terraced crop fields into ever higher and less stable mountain regions and to

extensive trampling of soil by uncontrolled tourism and increased numbers of grazing animals.

These impacts in turn trigger extensive erosion, flooding, landslides, mudslides, and avalanches that wipe out crops and livestock and kill humans. Ironically, the changing contours of upland areas also create problems such as silting and crop losses in the lowland areas from which many mountain residents migrated.

Jack Ives, professor of mountain geoecology at the University of Colorado in Boulder, described a recent trip to the Himalayas in northern India:

I visited the Darjeeling area in India just after 25 to 50 inches of rain had fallen during a three-day period. On the third day there were about 20,000 landslides in the Himalayas that killed some 30,000 people. And the 45-mile road from the plains up to Darjeeling at 7,500 feet was cut in 92 places. Rivers debouching from the mountain range had destroyed many rich farming and tea-growing areas by burying them under tons of sand and gravel.

ILLUSTRATIONS The tragedy of the commons is illustrated in Figures 99 to 105. In a natural undisturbed landscape, vegetation protects the soils and the environment (Figure 99). With farming and clearing (Figure 100), erosion is accelerated and runoff increases. At this stage careful soil conservation and soil management practices can achieve a sustainable production, but investment beyond the short run is usually necessary (with contributions to soil erosion control from society as a whole, beyond the capability of the individual farmer). When overcultivated and overgrazed (Figure 101), the productivity of the area has begun to decline and the yields of the soils are diminished. Continued overgrazing (Figures 102 and 103) will diminish the landscape to a wasteland (Figure 104). These trends are idealized in these figures, of course, and may require hundreds or thousands of years to reach the ultimate wasteland, but elements of the trends can be observed in nearly all landscapes. Terraces, contour cultivation, strip cropping, reforestation, fencing with rotational grazing, grass seedings, irrigation, drainage, and so on, are all good management practices that can help to achieve sustained productivity of landscape areas and reverse the process of the "tragedy of the commons." Overgrazing with goats (Figure 105), clearcutting of forests, burning of crop residues, and so on, are likely to be detrimental actions accelerating and contributing to lowering of productivity in the "tragedy of the commons." Once soil is eroded away, it can never be replaced. Page 107 of the textbook illustrates real situations applicable to Figures 99 to 105.

EROSION Soil erosion is a good example of the "tragedy of the commons," because it is unnoticed or ignored by most people. Yet the insidious incremental decline in productivity caused by soil erosion has serious consequences over the long run (Brink et al., 1977; Brown, 1978, 1981; Brown and Eckholm, 1974; Crittenden, 1980a, b; Faber, 1977; Flattau, 1983; Harnack, 1980; King, 1978; Risser, 1981). Soil erosion control measures are well known (see exercise on Erosion Control), but most are unprofitable in the short run. Each society has a stake in the long-run implications of erosion control, so that farmers and ranchers should be given technical and financial assistance to ensure the sustainable capability of the soils for production for future generations. Unfortunately, during periods of good weather and surplus production the problems are not realized. During periods of drought, floods, and scarcity it is too late to correct for the yield losses due to soil erosion. What is needed for the future is sustained planning for the long run to ensure good management and husbandry of the soils and the environment for future generations—and avoidance of the "tragedy of the commons."

EXAMPLES Many examples exist in every community to illustrate the "tragedy of the commons" —where arid regions are overgrazed and eroded; where parklands are trampled and compacted; where forests are clearcut without replanting; where urban development destroys the prime farmlands; where continuous row crops result in excessive soil losses; where soils and groundwaters are polluted by excessive dumping at waste disposal sites; or where drainage patterns are altered unknowingly in the uplands so that flood damage is caused in the lowlands. At Love Canal near Niagara Falls in New York State, excessive dumping of toxic wastes has caused gross pollution of the soils and the environment; the dumping was a result of the "tragedy of the commons" toward destruction of the environment to avoid more expensive disposal costs. In New York State each

year, more than 1,000 farms go out of business due to soil erosion, productivity declines, inadequate soil resource management, and other economic problems contributing to the "tragedy of the commons"; many of the farm failures could have been avoided with greater knowledge of the soil resources. In the western High Plains, from Montana to Texas, marginal soils are being plowed in order to realize quick profits, without thought to the severe consequences of wind and water erosion (Schmidt, 1982). In many cities, excessive groundwater pumping is causing great problems of soil subsidence (Reinhold, 1982); pollution is common where inadequate knowledge about soils and the environment is not used in planning and development (Blanc, 1980). In many places (Iverson et al., 1981), disregard for the soils and the environment causes many problems for people living in the areas.

In the tropics, many soils are in a fragile state of infertility and can be easily damaged by mismanagement to achieve "tragedy of the commons" situations (Janzen, 1973; Smith, 1981); careful management, however, can help to realize sustained and increased production on certain soils (Sanchez et al., 1982). In deserts, soils and the environment need special attention to reclaim areas where the "tragedy of the commons" has already caused ecological disruptions (Clawson et al., 1971; Evenari et al., 1971). Many problems in the world are caused by inadequate allocations and destructive uses of resources (Westing, 1980); many disasters (Hinds, 1983), such as floods, famines, and landslides, are caused by inadequate use and management of soils and the environment. Many have a gloomy view of the future (Carter, 1980), but better knowledge of our resources and better use of the resources could help to achieve a better world in the future (Farnsworth, 1982; Meadows et al., 1974).

PERSPECTIVES Students should be warned that different people have different perspectives relating to the "tragedy of the commons." Many executives of chemical companies have vested interests in cheap disposal of toxic wastes, and they are not likely to have great respect for soils and the environment. Fertilizer company stockholders are not likely to advocate low rates of chemical applications to soils to avoid environmental pollution (Olson et al., 1982). Both optimistic and pessimistic views are often presented on the same subject

(Sanchez et al., 1982; Smith, 1981). Environmentalists are sometimes emotional and uncompromising on ecological issues. Developers are often concerned only with a building site, and inconsiderate of the soils and the environment around it. Dairy farmers are often involved mostly with the health of their cows, and do not give equal attention to their soils. Students interpreting data relating to the "tragedy of the commons" should keep in mind all of the possible self-interests of the various people involved, and make interpretations objectively.

ASSIGNMENTS Students should make a list of all possible planning and development problems in their community and project areas relating to the concept of the "tragedy of the commons." Collect newspaper articles such as those cited here to show how political and economic decisions affect the community; sometimes it is necessary to "read between the lines" to get a correct understanding of situations. Construct hypothetical scenarios to predict what might happen if houses are built on floodplains, if subdivisions are put on steep unstable slopes, if erosive soils are intensively farmed, if clearcut timber areas are not replanted, if garbage landfill burial is located in soils with poor characteristics, if overgrazing is accelerated, and so on. All communities have problems of soil management, waste disposal, soil resource allocations, and so on—but many localities also have unique soil problems. This philosophical exercise helps to bring together all the details of the textbook and Field Guide, and make students, teachers, laypersons, and others think about long-range implications of decisions involving uses of soils and the environment.

REFERENCES

Anonymous. 1974. African drought. Time. 8 April. Pages 40–41.

Baden, J. 1980. Earth day reconsidered. The Heritage Foundation, Washington, DC. 108 pages.

Blanc, S. 1980. Toxic chemicals: Finding their way to the tap (from land to water). The Post-Standard (Syracuse, NY). 15 Sept. Pages A1 and A7.

Brink, R. A., J. W. Densmore, and G. A. Hill. 1977. Soil deterioration and the growing world demand for food. Science 197:625–630.

Brown, L. R. 1978. A biology lesson for economists. The New York Times. 9 July. Page 16F.

Brown, L. R. 1981. World population growth, soil erosion, and food security. Science 214:995–1002.

Brown, L. R. and E. P. Eckholm. 1974. Grim reaping: This year the whole world's short of grain. The New York Times. 15 Sept. Page 6.

Carter, L. J. 1980. Global 2000 report: Vision of a gloomy world. Science 209: 575–576.

Clawson, M., H. A. Landsberg, and L. T. Alexander. 1971. The agricultural potential of the Middle East. American Elsevier Publishing Company, Inc., New York. 312 pages and soil maps.

Coates, D. R. (Editor). 1972. Environmental geomorphology and landscape conservation, Vol I: Prior to 1900. Benchmark Papers in Geology. Dowden, Hutchinson, & Ross, Inc., Stroudsburg, PA. 485 pages.

Coates, D. R. (Editor). 1977. Landslides. Vol. III. Reviews in Engineering Geology. The Geological Society of America, Boulder CO. 278 pages.

Cowell, A. 1982. Africa's urban migration saps its agrarian strength. The New York Times. 19 Sept. Page 20E.

Crittenden, A. 1980a. Lack of U.S. funds cited in fight against erosion. The New York Times. 27 Oct. Pages D1 and D4.

Crittenden, A. 1980b. Soil erosion threatens U.S. farms' output. The New York Times. 26 Oct. Pages 1 and 55.

Crowe, B. L. 1969. The tragedy of the commons revisited. Science 166:1103–1107.

Eckholm, E. P. 1975. The deterioration of mountain environments: Ecological stress in the highlands of Asia, Latin America, and Africa takes a mounting social toll. Science 189:764–770.

Eckholm, E. P. 1976. Losing ground: Environmental stress and world food prospects. W. W. Norton and Company, Inc., New York. 223 pages.

Evenari, M., L. Shanan, and N. Tadmor. 1971. The Negev: The challenge of a desert. Harvard University Press, Cambridge, MA. 345 pages.

Faber, H. 1977. Albany is requiring the conservation of soil and water by landowners. The New York Times. 17 April. Page 51.

Farnsworth, C. H. 1982. The game people play over at the World Bank. The New York Times. 18 Oct. Page A12.

Franke, R. W. and B. H. Chasin. 1980. Seeds of famine: Ecological destruction and the development dilemma in the West African Sahel. Land Mark Studies. Allanheld, Osmun, Montclair, NJ, and Universe, NY. 268 pages.

Flattau, E. 1983. Some refugees seek environmental havens. The Ithaca Journal. April 2. Page 10.

Gentry, A. H.and J. Lopez-Parodi. 1980. Deforestation and increased flooding of the upper Amazon. Science 210: 1354–1356.

Hardin, G. 1968. The tragedy of the commons. Science 162:1243–1248.

Harnack, C. 1980. In Plymouth County, Iowa, the rich topsoils going fast. Alas. The New York Times. 11 July. Page A25.

Hinds, M. D. 1983. On watch for disasters across the U.S.

The New York Times. 8 Jan. Page 10.

Horsfall, J. G. 1972. Agricultural strategy in the tragedy of the commons. Agricultural Science Review 10:17–22.

Iverson, R. M., B. S. Hinckley, R. M. Webb, and B. Hallet. 1981. Physical effects of vehicular disturbances on arid landscapes. Science 212:915–917.

Janzen, D. H. 1973. Tropical agroecosystems. Science 182: 1212–1218.

King, S. S. 1978. Farms go down the river. The New York Times. 10 Dec. Page E9.

Leopold, A. S. 1962. The desert. Time-Life International, Weert, The Netherlands. 192 pages.

Love, G. A. and R. M. Love. 1970. Ecological crisis: Readings for survival. Harcourt Brace Jovanovich, Inc., New York. 342 pages.

Lowdermilk, W. C. 1978 (reprinted). Conquest of the land through 7,000 years. Agricultural Information Bulletin 99, Soil Conservation Service, U.S. Dept. of Agriculture, U.S. Government Printing Office, Washington, DC. 30 pages.

Meadows, D. L., W. W. Behrens III, D. H. Meadows, R. F. Naill, J. Randers, and E. K. O. Zahn. 1974. Dynamics of growth in a finite world. Wright-Allen, Cambridge, MA. 638 pages.

Olson, G. W. 1981. Archaeology: Lessons on future soil use. Journal of Soil and Water Conservation 36:261–264.

Olson, R. A., K. D. Frank, P. H. Grabouski, and G. W. Rehm. 1982. Economic and agronomic impacts of varied philosophies of soil testing. Agronomy Journal 74:492–499.

Reinhold, R. 1982. Houston's great thirst is sucking city down into the ground. The New York Times. 26 Sept. Page 28.

Riding, A. 1974. Haiti losing fight against erosion: Both cities and countryside feel its devastating effect. The New York Times. 23 June. Page 15.

Risser, J. 1981. A renewed threat of soil erosion: It's worse than the Dust Bowl. Smithsonian Magazine. 4 Mar. Pages 120–131.

Roos, L. L. Jr. 1971. The politics of ecosuicide. Holt, Rinehart and Winston, Inc., New York. 404 pages.

Sanchez, P. A., D. E. Bandy, J. H. Villachica, and J. J. Nicholaides. 1982. Amazon basin soils: Management for continuous crop production. Science 216:821–827.

Schmidt, W. E. 1982. Plowing of plains in the west stirs fear of new Dust Bowl. The New York Times. 14 May. Pages A1 and A18.

Smith, N. J. H. 1981. Colonization lessons from a tropical forest. Science 214:755–761.

Thomas, W. L. Jr. (Editor). 1965 (5th printing). Man's role in changing the face of the earth. University of Chicago Press, Chicago. 1193 pages.

Thomas, J. 1980. In northwestern Haiti, people are principal export. The New York Times. 26 Sept. Page A22.

Wade, N. 1974. Sahelian drought: No victory for western aid. Science 185:234–237.

Webster, B. 1982. Mountains worldwide imperiled as man and nature collide. The New York Times. 21 Dec. Pages C1 and C2.

Westing, A. H. 1980. Warfare in a fragile world: Military impact on the human environment. Stockholm International Peace Research Institute, Taylor & Francis Ltd., London. 250 pages.

FIGURE 99/*Natural forest vegetation on valleys, mountains, and hills. The trees protect the soil from erosion, and gradual infiltration of rainwater recharges the groundwater aquifers. Virgin land, even in dry climates, is able to support considerable vegetation if not disturbed. The roots of trees and plants secure the soil and hold water, thus preserving the area from erosion (adapted from Leopold, 1962).*

FIGURE 100/*Farming in the valleys and gradual clearing of the hills and mountains causes increased erosion and flooding. In arid regions, salts are leached into the seepage spots and deposited in stages of increasing salinity and alkalinity. Cultivation of the valleys and timber cutting on the slopes removes roots and bares the soils to wind and water erosion. Floods flush deposits of gravel from the upper slopes down onto the plain (adapted from Leopold, 1962).*

FIGURE 101/*Intensification in use of steep slopes degrades all the soils and the entire environment. Nearly level farmlands are degraded by sedimentation and salinization (in arid regions), and uplands are eroded. Runoff is greatly accelerated. Declining productivity of the flats may cause them to be abandoned to herds of cattle. Farmers gradually move up and up to steeper and steeper slopes, where the hazards of accelerated erosion of soils is much greater (adapted from Leopold, 1962).*

FIGURE 102/*Destruction of the forest reduces infiltration and increases erosion and runoff. Increased intensity of grazing with larger numbers of animals reduces and degrades the remaining vegetation. Lost fertility of the steeper hillsides caused by soil erosion and runoff renders them useless for further cultivation, and the processes of erosion by constant overgrazing is intensified as the cattle move up the slopes (adapted from Leopold, 1962).*

FIGURE 103/*The "tragedy of the commons" is evident when the productivity of the area is degraded to a low level. What was once productive in a harmonious relationship of soils and vegetation in the environment is now a wasteland populated only by underfed grazers. The last stages in the destruction of the already barren landscape occur when there is no longer enough browse for cattle, and the area is turned over to sheep and goats to be stripped clean (adapted from Leopold, 1962).*

FIGURE 104/*The "tragedy of the commons" is complete when the area serves no useful purpose. The exploitive human and animal populations have moved on, leaving wastelands in their wake. Total desolation of the once fertile region is now complete. Much of the original topsoil has disappeared and large sections of bedrock and infertile subsoils are exposed. Although the climate may have not changed, the increased runoff makes the region more arid. The dusty land can no longer support economic activity. The tragedy of the commons is bequeathed onto the future generations (adapted from Leopold, 1962).*

FIGURE 105/Goats are an indication of overgrazing of the soils of a society. They are voracious eaters, and agile in climbing steep slopes. Their appetites are enormous, and they will eat almost any vegetation and other items not considered edible by many other animals. Large herds of goats grazing freely over the landscape are often a sign of the "tragedy of the commons" in failure of soil conservation and soil management. Goats are popular in many low-income countries because they require little care and investment and return some profit; overpopulation and overgrazing by goats indicates disregard for the soils and the environment and is an open invitation to the "tragedy of the commons."

Strategic implications

PURPOSE Strategy is the utilization, during both peace and war, of all of a nation's resources, through large-scale, long-range planning and development, to ensure security or victory. Tactics, in contrast, deal with the use and deployment of troops in actual combat. Strategic implications in use of information about soils and the environment emphasize the extreme importance of resource allocations. Increasingly, the well-being of all nations has interlinkages involving use and abuse of resources. The purpose of this exercise is to encourage students, teachers, and laypersons to think broadly about the long-range implications involving uses of soils and the environment, hopefully to help develop a nation's security and harmonious relationships with other nations. The exercises "Land Uses and Soils" and the "Tragedy of the Commons" in this Field Guide and the textbook chapters on "Archaeological Considerations" (pages 130-142) and "Planning for the Future" (pages 143–160) should be studied for this exercise —along with the references cited here.

REVIEW Each student should select an area in which to do long-range strategic planning. The area selected can be a small project area, or a region, state, or country—or a larger region involving several countries (adjacent or separate). Soil information available may be detailed (at large scale for project area) or general (e.g., FAO World Soil Map for countries—see pages 59–61 and 143–148 of the textbook). Review all the available relevant information relating to landscapes, geomorphology, soils, and the environment. Make a list as comprehensive as possible of all the implications of different soil areas to past development and human occupation of the area. Then project future progress and development based on your knowledge of the soil conditions and several scenarios of events that might happen external to the area but nevertheless influential to the area in some way. Get ideas from the readings, and do not be afraid to use your imagination. Remember that human nature has not changed very much, and that many events of the past will probably happen again in the future. Processes of human greed, alliances, war, peace, development, destruction, progress,

regression, and so on, will probably operate in the future as in the past with some modifications for changing technology.

READINGS Readings on strategic implications in uses of soils and the environment should be broad and varied, with initiative and imagination in their selection. Newspapers, magazines, journals, books, and other publications should all be investigated. Newspapers and magazines are especially valuable to pick out important problems of the moment and project future trends; journals and books help to document historical trends and quantify past strategic implications of soil properties.

Herodotus (Selincourt, 1954) was one of the early observers of the strategic implications of soil properties. He was born about 490 B.C. in Greece, traveled widely for that time, and wrote the histories of the Persian Wars. He wrote

> The soil of Egypt does not resemble that of the neighbouring country of Arabia, or of Libya, or even of Syria...but is black and friable as one would expect of an alluvial soil formed of the silt brought down by the river from Ethiopia. The soil of Libya is, as we know, reddish and sandy, while in Arabia and Syria it has a larger proportion of stone and clay.

Herodotus was a keen observer of geomorphology, and wrote down his thoughts about erosion and sedimentation in various places. The invasion army of the Persian king Xerxes was transported by ships through a dug canal, and Herodotus wrote of some of the engineering difficulties involved in excavations of soils for the canal:

> I will now describe how the canal was cut. A line was drawn across the isthmus from Sane and the ground divided into sections for the men of the various nationalities to work on. When the trench reached a certain depth, the labourers at the bottom carried on with the digging and passed the soil up to others above them, who stood on ladders and passed it on to another lot, still higher up, until it reached the men at the top, who carried it away and dumped it. Most of the people engaged in the work made the cutting the same width at the

top as it was intended to be at the bottom, with the inevitable result that the sides kept falling in, and so doubled their labour. Indeed they all made this mistake except the Phoenicians who in this—as in all other practical matters—gave a signal example of their skill. They, in the section alloted to them, took out a trench double the width prescribed for the actual finished canal, and by digging at a slope gradually contracted it as they got further down, until at the bottom their section was the same width as the rest.

Preparations for the movement of the invasion army of Xerxes were elaborate according to Herodotus (Selincourt, 1954), and involved large supply requisitions from more fertile soil areas which were transported and stored at critical places along the invasion route:

> The greatest quantity was collected at a place called Leuce Acte in Thrace; other dumps were at Tyrodiza in Perinthian territory, Doriscus, Eion on the Strymon, and in Macedonia.

Strategic implications of soils and environmental resources are increasingly being recognized as important to world peace and relations between nations. Worldwide prosperity or depression is linked to the fate of each nation and environmental management (Colinvaux, 1980; Frisch, 1978; Hechinger, 1982; Lewis, 1983; McNeill, 1980; Shabecoff, 1982). Food surplus and scarcity are dependent on political and economic conditions, and unbalance in governmental management systems can be very damaging to large populations (Abelson, 1975a, b; Chancellor and Goss, 1976; Dando, 1980; Kupperman et al., 1975; Pimentel et al., 1976; Rensberger, 1974). In addition to area interdependence, solving problems also depends on cooperation of different disciplines. Saint and Coward (1977) have stated how components of various systems are oriented for agricultural production:

> Systems theory finds expression in an ecological systems approach in which both physical and organizational aspects of agricultural production are included as components of the same system: that is, the system perspectives extend beyond social organization to include agronomy (soils and crops), biology, and ecology, and vice versa. Relations are explored which link the environment, the crop, the crop producers, and the crop-producing community. Natural processes and social processes are seen as intertwined. Also, the relations between agriculture and the rest of society can be explored—for example, the impact of urbanization on agriculture.

The economics staff of the International Maize and Wheat Improvement Center (CIMMYT, 1981) devised a tabular display of the factors influencing corn production (Table 66). Practically all of the items listed under farmer circumstances and practices are influenced by soil conditions. Texture, slope, erosion, wetness, flooding, water-holding capacity, pH, rooting depth, watertable, hardpan, permeability, stoniness, rockiness, elevation, aspect, and so on, all influence priorities, allocations, and hazards (physical, biological, economic) associated with farming. The nature and timing of all practices (preparation, planting, weeding, harvesting, etc.) are also affected by soil properties. Agriculture, one of the most basic sectors of nations economies, is heavily dependent on the nature of soils and the environment. Management, improvement, and degradation of the soils play large roles in determining the prosperity or decline of a society.

Lester Brown (1975, 1977, 1981a, b) has written extensively on problems relating to agriculture and the soils. He argues effectively (Brown, 1977) that we need to redefine "national security" to include internal health and well-being of the people within a country and the related wise use of the natural resources, and not confine national security considerations to military expenditures. National security is considerably dependent on the soils and their careful management (Brown, 1977):

> Once topsoil is lost, a vital capacity to sustain life is diminished. With soil as with many other resources, humanity is beginning to ask more of the earth than it can give.

> More and more the "carrying capacities" of biological systems are being ignored and exceeded.

> In many Third World countries population growth is now acting as a double-edged sword, simultaneously expanding demands on the biological systems while destroying the resource bases.

> Closely related to the contribution of population growth to food insecurity in the Third World is a complex of negative ecological trends

—deforestation, overgrazing, desert encroachment, soil erosion, and flooding.

The unfolding stresses in this relationship initially manifest themselves as ecological stresses and resource scarcities. Later they translate into economic stresses—inflation, unemployment, capital scarcity, and monetary instability. Ultimately, these economic stresses convert into social unrest and political instability.

National security requires a stable economy with assured supplies of materials.... In this sense, frugality and conservation of materials are essential to our national security. Security means more than safety from hostile attack; it includes the preservation of a system of civilization.

Strategic implications of soils and the environment were important in the past (Adams et al., 1981) and will continue to be important in the future (CEQ, 1980). George Washington is reputed to have favored Washington, D.C., as a site for the capital of the United States partly because of the good soils in the vicinity (Molotsky, 1982). Common knowledge is that good soils are beneficial to human health (Davies, 1977), that poor soils foster poverty cycles (Charney et al., 1975; Freis, 1981; Hogue, 1983; Nordheimer, 1983), that many diseases are related to the soils (Lindsey, 1981; Page et al., 1976; Rohl et al., 1982), and that soil abuse is somehow harmful to society (Lindsey, 1983; Traver, 1982; Walsh, 1977).

Many regional development programs are linked to soil areas. Erosion problems in the northwestern United States are being attacked in comprehensive research programs (Oldenstadt et al., 1982); Appalachia poverty was reduced through soil inventory, management, and developmental efforts (Anonymous, 1981); future programs will likely be guided increasingly by new technology applied to alleviating soils problems (Adrien and Baumgardner, 1977; Petersen and Beatty, 1981). Soil erosion is one example of a worldwide problem that needs constant attention for management and improvement of soils and environmental resources (Eckholm, 1976; El-Swaify et al., 1982; FAO, 1979; Young, 1976).

Some of the strategic implications of soils and the environment are extremely complex. When economic times are difficult for farmers (Herbers, 1982; Schmidt, 1983), other segments of society also suffer (Scott, 1982). Statistics indicate that 1,000 farms go out of business each year in New York State (Cornell University, 1981), and New York farm productivity lags behind other states (Mundell, 1982). Questions that must be asked in the future include:

1. Do strategic implications require a strong agriculture in the United States (or in any country)?
2. Does agriculture in New York State have strategic implications in comparison with other states?
3. Do certain areas within New York State need governmental assistance in soil conservation and management programs?
4. Does erosion control have strategic implications for future productivity?
5. What are the best procedures and programs to ensure soil protection and productivity?

IRAN EXAMPLE The Iranian situation is not understood very well (Wade, 1979). Ironically, the strategic implications of soils and desertification is more responsible for the plight of the people than any social or political manipulations, especially in the long run (Olson, 1980). Abuse of the land in Iran provides classic illustrations of the "tragedy of the commons" (Hardin, 1968). We should carefully study the environment and the past human abuse of land in Iran, because these extreme examples will be increasingly a part of the strategic scenarios in other nations of the region and elsewhere (Clawson et al., 1971; Evenari et al., 1971; Olson and Hanfmann, 1971).

Generally, Iran is a geomorphic bowl (Fisher, 1968) with an outer mountain rim surrounding an irregular lower (but not low-lying) interior. The highland erosion causes extensive fans to form in the upper reaches. Gradually, the rate of erosion diminishes downslope to areas of alluviation and deposition, where finer-textured soils and lower gradients occur. At the core of the typical enclosed endoreic basins (kavirs) are salty clay deposits and intermittent salt lakes. In the arid climate, mineral deposits form giant aureole in a regular stratification of soils in the landscapes: at the lower levels, calcium, magnesium, and sodium chlorides occupy basin floors; on the next level, sulfates of these three elements are found; and carbonates form an upper layer bordering the hills. Land use conforms rigidly to the salt and soil zones, and changes from grazing in the mountains to dryland farming in the fans of the foothills to intensive irrigation agriculture, and finally culminates in sparse grazing in salty clay flats or intermittent salt lakes in the cores of the kavirs. Ancient underground aqueduct systems (qanats) tap the fresh water in the upper gravelly aquifers, and lead it out for irrigation downslope onto finer-textured soils. Even in the capital city of Tehran the quality of urban life and habitations is closely related to the soils and to the environment: elite residential units occupy the footslopes of the mountains, and the buildings steadily deteriorate to the slums at the edge of the salty clay flats of the desert.

Much of the history of soil use in Iran relates to the hostility of the geomorphic setting. Absence of water has dictated that control of much of the land falls to rich absentee landlords who have the only available capital to develop water systems. Despotic rulers with desires for luxurious living have helped to keep much of the rural population in poverty through the centuries (Fisher, 1968). Earthquakes are common in many areas, and have destroyed thousands of people and houses in countless villages. Floods, droughts, invasions, epidemics, sieges, and locust plagues are part of the typical experience of Iranian settlements. The history of Nishapur (home of the poet Omar Khayyam) serves as an example of the hostility of the Iranian environment and experience. It was apparently built in the third century, developed and expanded into the eleventh century, was sacked by Turkish hoards in 1145, and damaged by earthquakes in 1145 and 1208. In April 1221, the Mongols exterminated the population, killing even the dogs and cats, and in token of the destruction they leveled the ruins, even plowing and sowing the soil (Wagret, 1972). There was a partial recovery, because a number of survivors were recorded after the earthquake of 1267. The town was rebuilt, and again declined. In the eighteenth century the Afghan invasion, followed by Turcoman raids, reduced the population even further. More than 250,000 archaeological sites in Iran mark places where similar events are recorded by artifacts buried in the soils (Matheson, 1973). In many places where sites were abandoned, the most critical factor in the disappearing of the population was destruction of the water system.

The general process of desertification, human degradation of the soils, and illustration of the "tragedy of the commons" in Iran in this geomorphic environment has been excellently described (Smith, 1966) using the desert around Kirman as an example:

> The center of the plain of Kirman is, except for small settlements, quite barren. The rest is hard desert and dry sand. This sand is a menace, for the winds blow it about. Initially, only the crops are ruined; but if the wind blows from the same direction for long enough, then a village may become submerged. Such remains can be seen scattered over the plain. This danger is recent, is increasing and the fault of the Kirmanis themselves. Under normal conditions most deserts will become static: small plants manage to grow and so help to bind the sand particles; then larger plants survive and the sand around them becomes harder and acts as a protective crust. Rain followed by sun increases the firmness of this surface layer and, if left to themselves, small areas of desert do not constitute an engulfing menace. Although the livelihood of the whole basin depends upon the stability of the land, all sizeable plants are uprooted to be burnt as fuel....Steadily the firmness is being lost; each year more of the desert begins to move. Great walls are built to arrest the advancing sand, but whenever there is a wind the sand increases its attack; the outlying fields are covered, the dust is spread over the town, the roads become blocked and have to be cleared. Nothing is done to prevent the problem. It is nobody's affair. It is the will of Allah if great winds are sent....The desert belongs to no one; ownership applies only to land which is cultivated. No responsibility is involved, so that their future welfare and the problem is allowed to rest until the future.

The qanat systems of underground aqueducts provide about 75 percent of the water used in Iran (Bybordi, 1974). Total length of the qanat tunnels is estimated at 350,000 km. The "wells" used to dig and maintain the tunnels are hand dug and spaced about 50 meters apart. The deepest wells are generally not more than 100 meters deep, although in rare cases they may be more than 300 meters deep. The longest known qanat is about 80 km in length (Smith, 1966).

As overgrazing and vegetative destruction continues, of course, erosion increases and salinization becomes a greater problem. Accelerated runoff lowers water tables so that qanat systems become worthless or their flows are reduced. Conservation of soils and vegetation could solve many of the problems of the environment, but few trends are discernible to reverse the soil degradation in the tragedy of the commons. More human unrest and misery can be expected in Iran and elsewhere where soils (the most valuable resource) are abused and misused. People are a reflection of their ecological habitat more than they realize.

ASSIGNMENT In accord with the ideas, concepts, and readings presented about the strategic implications of soils and the environment, prepare a report to illustrate the importance of soils to large-scale, long-range planning and development strategy for a nation, region, or area. State past and present problems in soil and environmental management, and construct several scenarios of events that might happen in the future to influence coming trends. Outline programs that could encourage better use and conservation of soils and the environment for the next generations. Discuss implications of the "tragedy of the commons" and "intergenerational equity" (where future generations have an equity in present existing resources, and the present generation has no right to exploit or abuse those resources excessively). Conclude by summarizing effects of present trends, programs, and legislation, and suggest or recommend changes for the future.

REFERENCES

Abelson, P. H. (Editor). 1975a. Food. Special Issue. Science 188:502–654.

Abelson, P. H. 1975b. The world's disparate food supplies. Science 187:1.

Adams, R. E. W., W. E. Brown, Jr., and T. P. Culbert. 1981. Radar mapping, archaeology, and ancient Maya land use. Science 213:1457–1463.

Adrien, P. M. and M. F. Baumgardner. 1977. Landsat, computers, and development projects. Science 198:466–470.

Anonymous. 1981. The costly federal effort to save Appalachia: Was $15 billion well spent? The New York Times. 27 Sept. Pages 1 and 32.

Brown, L. R. 1975. The world food prospect. Science 190:1053–1059.

Brown, L. R. 1977. Redefining national security. Worldwatch Paper 14, Worldwatch Institute, Washington, DC. 46 pages.

Brown, L. R. 1981a. An enduring society. The New York Times. 19 Dec. Page 15.

Brown, L. R. 1981b. World population growth, soil erosion, and food security. Science 214:995–1002.

Bybordi, M. 1974. Qanats of Iran: Drainage of sloping aquifers. Proceedings of the American Society of Civil Engineers, Journal of the Irrigation and Drainage Division 100:245–253.

CEQ. 1980. The global 2000 report to the President. Vol. I: The summary report (50 pages), and Vol. II: The technical report (766 pages). Council on Environmental Quality and Dept. of State, U.S. Government Printing Office, Washington, DC.

Chancellor, W. J. and J. R. Goss. 1976. Balancing energy and food production, 1975–2000. Science 192:213–218.

Charney, J., P. H. Stone, and W. J. Quirk. 1975. Drought in the Sahara: A biogeophysical feedback mechanism. Science 187:434–435.

CIMMYT. 1981. Assessing farmers' needs in designing agricultural technology. Occasional Paper, International Agricultural Development Service, New York. 12 pages.

Clawson, M., H. A. Landsberg, and L. T. Alexander. 1971. The agricultural potential of the Middle East. American Elsevier Publishing Company, Inc., New York. 312 pages and soil maps.

Colinvaux, P. 1980. The fates of nations: A biological theory of history. Simon and Schuster, New York. 384 pages.

Cornell University. 1981. Facts and figures. New York State College of Agriculture and Life Sciences, Cornell University, Ithaca, NY. 38 pages.

Dando, W. A. 1980. The geography of famine. Halsted Press (John Wiley), Somerset, NJ. 365 pages.

Davies, D. M. 1977. Is soil secret of longevity? The Ithaca Journal. 30 Sept. Page 24.

Eckholm, E. P. 1976. Losing ground: Environmental stress and world food prospects. W. W. Norton and Company, Inc., New York. 223 pages.

El-Swaify, S. A., W. Kussow, and J. Mannering (Editors). 1982. Soil erosion and conservation in the tropics. Special Publication 43, American Society of Agronomy, Madison, WI. 149 pages.

Evenari, M., L. Shanan, and N. Tadmor. 1971. The Negev: The challenge of a desert. Harvard University Press, Cambridge, MA. 345 pages.

FAO. 1979. If the land dies....259 photos with text and filmstrip. Food and Agriculture Organization of the United Nations, Rome, Italy. 70 pages.

Fisher, W. B. 1968. The land of Iran: Vol. I of the Cambridge history of Iran. Cambridge University Press, London. 784 pages.

Freis, R. 1981. Soil fosters poverty cycle. The Culpeper News (VA). 15 Jan. Pages 1 and 2.

Frisch, R. E. 1978. Population, food intake, and fertility. Science 199:22–30.

Hardin, G. 1968. The tragedy of the commons. Science 162:1243–1248.

Hechinger, F. M. 1982. The world view back in favor. The New York Times. 26 Oct. Page C5.

Herbers, J. 1982. Recession dimming future of family farms. The New York Times. 16 Oct. Pages 1 and 6.

Hogue, W. 1983. Brazil's Northeast is wilting under a smiling sky. The New York Times. 18 Jan. Page A2.

Kupperman, R. H., R. H. Wilcox, and H. A. Smith. 1975. Crisis management: Some opportunities. Science 187: 404–410.

Lewis, P. 1983. New interest in world growth. The New York Times. 19 Jan. Page D13.

Lindsey, R. 1981. Californians fight a curious regional disease. 27 Nov. Page A20.

Lindsey, R. 1983. Hard days cloud fields of San Joaquin valley. The New York Times. 20 Jan. Page A18.

Matheson, S. A. 1973. Persia: An archaeological guide. Noyes Press, Park Ridge, NJ. 330 pages.

McNeill, W. H. 1980. The human condition: An ecological and historical view. Princeton University Press, Princeton, NJ. 82 pages.

Molotsky, I. 1982. George Washington on real estate. The New York Times. 14 July. Page A20.

Mundell, H. 1982. Study claims New York farm productivity lags behind others. The Ithaca Journal. 21 Aug. Page 3.

Nordheimer, J. 1983. Again, for Highland Scots, the specter of eviction. The New York Times. 19 Jan. Page A2.

Oldenstadt, D. L., R. E. Allan, G. W. Bruehl, D. A. Dillman, E. L. Michalson, R. I. Papendick, and D. J. Rydrych. 1982. Solutions to environmental and economic problems (STEEP). Science 217:904–909.

Olson, G. W. 1980. The strategic situation in Iran according to the soils. Soil Survey Horizons 21:21–24.

Olson, G. W. and G. M. A. Hanfmann. 1971. Some implications of soils for civilizations. New York's Food and Life Sciences Quarterly 4:11–14.

Page, T., R. H. Harris, and S. S. Epstein. 1976. Drinking water and cancer mortality in Louisiana. Science 193: 55–57.

Petersen, G. and M. Beatty. 1981. Planning future land uses. Special Publication 42, American Society of Agronomy, Madison, WI. 71 pages.

Pimentel, D., E. C. Terhune, R. Dyson-Hudson, R. Rochereau, R. Samis, E. A. Smith, D. Denman, D. Reifschneider, and M. Shepard. 1976. Land degradation: Effects on food and energy resources. Science 194: 149–155.

Rensberger, B. 1974. 32 nations close to starvation. The New York Times. 2 Oct. Page 4E.

Rohl, A. N., A. M. Langer, G. Moncure, I. J. Selikoff, and A. Fischbein. 1982. Endemic pleural disease associated with exposure to mixed fibrous dust in Turkey. Science 216:518–520.

Saint, W. S. and E. W. Coward, Jr. 1977. Agriculture and behavioral science: Emerging orientations. Science 197:733–737.

Schmidt, W. E. 1983. In farm belt, fear of foreclosures rises. The New York Times. 16 Jan. Pages 1 and 20.

Scott, L. 1982. State health chief says more people are going hungry. The Ithaca Journal. 5 June. Page 3.

Selincourt, A. D. 1954. Herodotus: The histories. Penguin Books, Baltimore, MD. 624 pages.

Shabecoff, P. 1982. Group will study the environment: MacArthur Foundation grants $15 million to inaugurate Global Policy Institute. The New York Times. 4 June. Page A21.

Smith, A. 1966. Blind white fish in Persia. Unwin Books, London. 207 pages.

Traver, N. 1982. Eastern Colorado farmers fear return of the "Dust Bowl." The Ithaca Journal. 19 Mar. Page 14.

Wade, N. 1979. Iran and America: The failure of understanding. Science 206:1281–1283.

Wagret, P. 1972. Nagel's guide to Iran. Nagel Publishers, Geneva, Switzerland. 392 pages.

Walsh, J. 1977. Seveso: The questions persist where dioxin created a wasteland. Science 197:1064–1067.

Young, A. 1976. Tropical soils and soil survey. Cambridge University Press, Cambridge, England. 468 pages.

TABLE 66/*Inventory of potential influences on small farmer management practices for maize (adapted from CIMMYT, 1981).*

PRACTICES 1–25	SUBSISTENCE PRIORITIES		RESOURCE ALLOCATION			HAZARD AVOIDANCE — PHYSICAL			HAZARD AVOIDANCE — BIOLOGICAL			HAZARD AVOIDANCE — ECONOMIC		
	Preferred food staple or relish	Food needs at specific time of year	Land scarcity	Labor scarcity	Capital scarcity or poor cash flow	Rainfall uncertain	Floods	Soil erosion and degradation	Pests	Diseases	Weeds	Poor retail distribution	Food and cash crop prices uncertain	Seasonal variation in food prices
	A	B	C	D	E	F	G	H	I	J	K	L	M	N
A. ON THE TARGET CROP														
1. Choice of soil type		⊗	X	X		X	X	X		X				
2. Choice of location		⊗				X	X	X	X	X	X			
3. Methods of seedbed preparation				⊗	X	X		X			X			
4. Time of planting		⊗	X	⊗		X	X		X		X			X
5. Method of planting				X							X			
6. Spacing			X	X		X					X			
7. Plant population					X	X	X		X	X	X			
8. Variety used	⊗	⊗			X	X			⊗	X	X			
9. Number of plantings made	X	⊗		⊗		X	X		⊗					
10. Intercropping		X	X	⊗		X		X	X		X			
11. Relay cropping		X	X	X				X	X		X			
12. Frequency and timing of weeding				⊗	⊗			X	X	X	⊗			
13. Use of fertilizer			X		X	X		⊗			X	X		
14. Method and time of harvest	X	⊗	X	X					X	X				
15. Method of processing & storage		X			X				⊗					X
16. Use of herbicide				X	X						X	X		
17. Use of insecticides					X				X			X		
B. IN THE FARMING SYSTEM														
18. Growing preferred foods	⊗	X										⊗	X	X
19. Growing non-preferred foods		X	X	X		⊗			⊗	X		X		X
20. Crop rotation								⊗	X	X				
21. Renting of land			X											
22. Hire of labor and machines				X	X	X								
23. Labor reciprocity				X	X									
24. Winter land preparation				X	X	X		X	X		X			
25. Firing or bush fallow								⊗			X			

X Farmer circumstances potentially influencing choice of a management practice

⊗ Circumstances governing choice of practices in one study area.

Military campaigns

PURPOSE The ultimate causes of most human problems in the world are greed and misappropriation (improper allocations and mismanagement) of land and soil resources. In military campaigns and tactics most civilian interpretations of soil surveys are also applicable to military situations. Thus, ratings of soils for agricultural production are equally relevant to food supplies in peace and war; trafficability ratings of soils can be applied to farm tractors or Jeeps and tanks; and engineering ratings of soils for construction of airfields, roads, buildings, excavations, embankments, and so on, are applicable to prosperous or troubled times. Soil survey interpretations are vital also for complex societal decisions relating to all aspects of long-range strategic planning: military toxic and radioactive wastes, for example, may cause cancer and other diseases in civilian populations from improper disposal in soils many years after the burials have taken place (e.g., Love Canal in New York State, Times Beach in Missouri). Excessive soil erosion may cause yield declines, desertification, and other environmental problems contributing to mass migrations and refugee dislocations of populations with resultant social strife. Inadequate and improper land-tax and land-tenure systems are often causes for revolutions and wars. Seizures of territory belonging to other countries are usually connected to land pressures and real or perceived resource shortages. Hopefully, in the future, better understanding of the tragedy of the commons and strategic implications will reduce uses of soils for military campaigns. When rational judgments fail, military campaigns often result. This exercise is designed to encourage strategic thinking in the minds of citizens—students, teachers, and laypersons—in the hope that future wars can be avoided. If the lessons of history are not learned in avoidance of war, this information will be directly related to future military campaigns and tactics. Pages 46–48, 91–95, 101–104, 113–114, 140–142, and 143–160 of the textbook should be studied in preparation for this exercise. This section on "Military Campaigns" sequentially follows the Field Guide exercises "Tragedy of the Commons" and "Strategic Implications" in the uses of soils information.

READINGS The literature relevant to soil survey interpretations and military campaigns is extremely varied and extensive, and students are encouraged to read as widely as possible into the subject. In the broadest sense, of course, literature on the political, economic, social, geographic, and anthropologic causes of war is relevant to this exercise. Much of the technical literature is classified (see pages 143–148 of the textbook), but a great deal of information is also freely available to anyone searching for references on the subject. Of extreme importance are readings about historical aspects of soil use and abuse, and newspaper articles that help to predict future trends. The following references illustrate the kinds of readings that contribute toward achieving an understanding of different environments and human adjustments and maladjustments to them:

"The Indochina story" (Ackerman et al., 1970)
"Famine again perils the Sub-Saharan region" (Anonymous, 1983a)
"Shifting priorities" (Anonymous, 1983b)
"British said to take ridge close to Stanley airfield" (Apple, 1982)
"The wonder that was India" (Basham, 1954)
"Israeli Premier visits captured castle: Crusaders built the castle" (Chira, 1982)
"The Middle East: A physical, social, and regional geography" (Fisher, 1963)
"Soil geography and land use" (Foth and Schafer, 1980)
"Viet Nam: History, documents, and opinions on a major world crisis" (Gettleman, 1965)
"Inside Africa" (Gunther, 1955)
"Central Asia" (Hambly, 1969)
"Contemporary Latin America: A short history" (Hanke, 1968)
"Violence seems to grow in India's troubled state" (Hazarika, 1983)
"Transcript of statements at State Department of the military buildup in Nicaragua" (Inman, 1982)
"A matter of survival: Whoever controls the strategic Judea and Samaria controls Israel" (Katz, 1983)
"Political and social philosophy: Traditional and

contemporary readings" (King and McGilvray, 1973)

"Food sales by U.S. called peace tool: Report regards the Russians' reliance on American grain as a nuclear deterrent" (King, 1982)

"The pursuit of power: Technology, armed force, and society since A.D. 1000" (McNeill, 1983)

"The voice of Asia" (Michener, 1951)

"Forty years ago, the battle for North Africa" (Middleton, 1982a)

"British victory: Coordination and professionalism" (Middleton, 1982b)

"A short history of India" (Moreland and Chatterjee, 1962)

"Inventory and nomenclature of Viet Nam soils" (Nguyen Hoai Van, 1962)

"Development reconsidered" (Owens and Shaw, 1972)

"Conquest of Mexico" (Prescott, 1934)

"Determinants of soil loss tolerance" (Schmidt et al., 1982)

"A history of land use in arid regions" (Stamp, 1965)

"Man's role in changing the face of the earth" (Thomas et al., 1965)

"The Times concise atlas of world history" (Times, 1982)

"World of the Maya" (Von Hagen, 1964)

BIBLE The Bible (Metzger, 1982) contains many references to soils, relating to strategy and military campaigns and social problems. The Lord said to Moses

> Send men to spy out the land of Canaan....
>
> Go up into the hill country, and see what the land is, whether it is good or bad, rich or poor, whether the people who dwell in it are strong or weak, few or many, and whether their cities are camps or strongholds. Be of good courage, and bring back some of the fruit of the land.
>
> Every one then who hears these words of mine and does them will be like a wise man who built his house upon the rock; and the rain fell, and the floods came, and the winds blew and beat upon that house, but it did not fall, because it had been founded on the rock. And every one who hears these words of mine and does not do them will be like a foolish man who built his house upon the sand; and the rain fell, and the floods came, and the winds blew and beat against that house, and it fell; and great was the fall of it.

> A sower went out to sow. And some seed fell along the path, and the birds came and devoured it. Other seed fell on rocky ground, where it had not much soil, and immediately it sprang up, and when the sun rose it was scorched, and since it had no root it withered away. Other seed fell among thorns and the thorns choked it, and it yielded no grain. And other seeds fell into good soil and brought forth grain, yielding thirtyfold and sixtyfold and a hundredfold.
>
> Be confounded, O tillers of the soil, for the wheat and the barley have perished.
>
> The seed shrivels under the clods; the storehouses are desolate.
>
> The earth quakes before them; the heavens tremble.
>
> Behold, I am sending you grain, wine, and oil.
>
> ...hid in the caves and among the rocks of the mountains.

To the Church at Sardis (see pages 136–139 of the textbook; Book of Revelation): "I know your works; you have the name of being alive, and you are dead. Awake, and strengthen what remains. Remember what you received, and repent. If you will not awake, I will come like a thief, and you will not know at what hour...." Later, invasions, droughts, earthquakes, famines, fires, floods, landslides, sieges, and environmental degradation destroyed the city and the church at Sardis. The people did not awake, and the church at Sardis is no more. Today only a few Moslems inhabit the area, and the splendor of ancient Sardis is reflected only in the artifacts and in the archaeological remains. Much still lies buried beneath the soils.

REVOLUTIONS Many countries have severe social problems arising from misappropriation of soil resources and bad land-tenure systems. Revolutions will increasingly result if people are deprived of the necessities of life, especially if a few greedy elite amass most of the wealth for themselves. A dictator and a few wealthy families in a country, for example, may own most of the land, including the best soils, while most poor people are landless or functioning only as peasants farming small plots on steep and stony soils. The rich often use the best soils for pasture or export crops, which produce the greatest profits with minimal inputs. Often the best soils are merely appropriated by the rich few for speculative purposes. And the peasants are forced into poverty and hunger on the poorest of soils, where erosion and environmental degradation are likely to be accelerated. Consequent migrations of rural people to urban slums cause increasing discontent, especially when the urban wealth is highly visible and vividly contrasting to the squalor of the common people. The inventory and management of soils will be increasingly critical in preventing revolutions; land tenure and taxation systems must be equitable and reasonable in stable societies. Democratic ideals, socialism, communism, and other political, religious, and social philosophies will be competing to construct human systems that will persist in different countries. If democracy is to survive, social inequalities cannot be too great when each person has an equal vote. In the Communist Manifesto, Karl Marx places a great emphasis on land tenure and soil management (King and McGilvray, 1973):

> The proletariat will use its political supremacy to wrest, by degrees, all capital from the bourgeoisie, to centralize all instruments of production in the hands of the State (i.e., the proletariat organized as the ruling class), and to increase the total productive forces as rapidly as possible.
>
> These measures will of course be different in different countries.
>
> Nevertheless in the most advanced countries, the following will be pretty generally applicable:
>
> 1. Abolition of property in land and application of all rents of land to public purposes.
> 2. A heavy progressive or graduated income tax.
> 3. Abolition of all right of inheritance.
> 4. Confiscation of the property of all immigrants and rebels.
> 5. Centralization of credit in the hands of the State, by means of a national bank with State capital and an exclusive monopoly.
> 6. Centralization of the means of communication and transport in the hands of the State.
> 7. Extension of factories and instruments of production owned by the State; cultivation of waste lands; and improvement of the soil generally in accordance with a common plan.
> 8. Equal liability of all to labor. Establishment of industrial armies, especially for agriculture.
> 9. Combination of agriculture with manufacturing industries; and gradual abolition of the distinction between town and country by a more equable distribution of the population over the country.
> 10. Free education for all children in public schools. Abolition of childrens' factory labor in its present form. Combination of education with industrial production, etc.

VIET NAM Military campaigns in Viet Nam were highly dependent on the soils and their behavior (Nguyen Hoai Van, 1962). The people who best understood the local environments were often the victors in battles. Especially critical were determinations of soil trafficability along supply routes, and military occupations of areas with productive potential for agriculture. Something of the misery of the people in the military campaigns is expressed in this poem (Ackerman et al., 1970):

> On this land
> where each blade of grass is human hair
> each foot of soil is human flesh
> where it rains blood
> hails bones
> life must flower
>
> NGO VINH LONG

The ultimate cause of the recent Viet Nam conflicts, of course, relates back to the long history of competition for land areas and invasions and occupations over centuries. In 1924, Ho Chi Minh wrote on the condition of the peasants in Viet Nam (Gettleman, 1965):

> In former times, under the Annamese regime, lands were classified into several categories according to their capacity for production. Taxes were based on this classification. Under the present colonial regime, all this has changed. When money is wanted, the French Administration simply has the categories modified. With a stroke of their magic pen, they have transformed poor land into fertile land, and the Annamese peasant is obliged to pay more in taxes on his fields than they can yield him.

EUROPE General Dwight David Eisenhower has many references to "the bottomless mud" in his book *Crusade in Europe* (Eisenhower, 1948) about World War II military campaigns. Rainy weather conditions would often saturate soils with poor trafficability characteristics, and troops and vehicles "were hopelessly bogged down" (Figure 106). The frustration is reflected in numerous quotations of this nature:

> The rain fell constantly. We went out personally to inspect the countryside over which the troops would have to advance, and while doing so I observed an incident which, as much as anything else, I think, convinced me of the hopelessness of an attack. About thirty feet off the road, in a field that appeared to be covered with winter wheat, a motorcycle had become stuck in the mud. Four soldiers were struggling to extricate it but in spite of their most strenuous efforts succeeded only in getting themselves mired into the sticky clay. They finally had to give up the attempt and left the motorcycle more deeply bogged down than when they started.
>
> ...as winter approached, the winding roads leading into my little camp at Reims at times became impassable. One afternoon I was bogged down for three hours while waiting for a tractor to pull my car out of a ditch.
>
> It made satisfactory initial gains but the troops quickly found themselves involved in a quagmire of flooded and muddy ground and pitted against heavy resistance.

The importance of strategic location of soil map units is illustrated in this quote:

> The enemy's invariable practice upon capture of a hill or other feature was to plant his mines instantly, install his machine guns, and locate troops in nearby reserve where they could operate effectively against any force that we might send against them....

Regarding the location of forward airfields, the best was at Thelepte:

> It lay in a sandy plain, and operation from it were never interrupted by rain; only the occasional sandstorm impeded its use. Because of the advantages of this airfield, we placed on it large air formations and comparable quantities of supplies and repair facilities....

Terrain difficulties in mounting an attack are described in this quotation:

> Topographically Pantelleria presented almost dismaying obstacles to an assault. Its terrain was entirely unsuited to the use of airborne troops, while its coast line was so rocky that

only through the mouth of the island's one tiny harbor was it possible to land troops from assault boats....

Troop health (related to soil conditions) was of constant concern:

> The plain was infested with malaria. In no other area during the Mediterranean campaign did we suffer equal percentage losses from disease. At other points in Sicily we likewise had a serious casualty list from malaria, but Catania was the pesthole of the region.

A letter from General Eisenhower to General Marshall stated:

> I should like to point out that the so-called "good ground" in northern Germany is not really good at this time of year. That region is not only badly cut up with waterways, but in it the ground during this part of the year is very wet and not so favorable for rapid movement as is the higher plateau over which I am preparing to launch the main effort.

Even soil fertility was a major consideration in strategy:

> I still refused to consider a major offensive into the country (Holland). Not only would great additional destruction and suffering have resulted but the enemy's opening of dikes would further have flooded the country and destroyed much of its fertility for years to come. I warned General Blaskowitz, the German commander in Holland, to refrain from opening any more dikes and pointed out to him that nothing he could do in Holland would impede the speedy collapse of Germany.

Finally:

> From the beginning of the conquest of Sicily we had been engaged in a new type of task, that of providing government for a conquered population. Specially trained "civil affairs officers," some American, some British, accompanied the assault forces and continuously pushed forward to take over from combat troops the essential task of controlling the civil population.
>
> The American contingent had been trained in the school established at Charlottesville, Virginia. Later, groups of both British and American military government officers received further training in North Africa. They operated

under the general supervision of a special section of my headquarters.

Public health, conduct, sanitation, agriculture, industry, transport, and a hundred other activities, all normal to community life, were supervised and directed by these officers. Their task was difficult but vastly important, not merely from a humanitarian viewpoint, but to the success of our armies. Every command needs peace and order in its rear; otherwise it must detach units to preserve signal and road communications, protect dumps and convoys, and suppress underground activity.

MEXICO The *Conquest of Mexico* during 1516–1518 (Prescott, 1934), as recorded by Spanish participants, contains many references to soils in military campaigns. Cortes prepared for the campaign by assembling provisions from royal farms in Spain and other sources in Cuba. At the mouth of the Rio de Tabasco in Mexico, sand accumulations forced the general to embark in small boats with only part of his forces. Mangrove swamps were difficult to traverse. In initial encounters with unfriendly Indians, troop movements were greatly retarded "by the broken nature of the ground." At Vera Cruz: "It was a wide and level plain, except where the sand had been drifted into hillocks by the perpetual blowing of the norte (wind). On these sand hills he mounted his little battery of guns, so as to give him the command of the country." In a military display to intimidate the Indians, "he ordered out the cavalry on the beach, the wet sands of which offered a firm footing for the horses." At an encampment "the soldiers suffered greatly from the inconveniences of their position amidst burning sands and the pestilent effluvia of the neighboring marshes, while the venomous insects of these hot regions left them no repose, day or night." For the first colony in New Spain, a spot was selected for the new city "in a wide and fruitful plain, affording a tolerable haven for the shipping."

Difficulties were encountered "over the roads made nearly impassable by the summer rains," and higher elevations were considered to be "above the deadly influence of the vomito." Landscape descriptions are vivid as the army traveled across the Sierra Madre, the Tierra Caliente, volcanic terrain, and into Sierra del Agua. Strategic areas for agricultural production were identified for future occupation:

"The temperate climate of the tableland fur-

nished the ready means for distant traffic. The fruitfulness of the soil was indicated by the name of the country—Tlascala signifying the "land of bread." Its wide plains to the slopes of its rocky hills, waved with yellow harvests of maize, and with the bountiful maguey, a plant which, as we have seen, supplied the materials for some important fabrics. With those, as well as the products of agricultural industry, the merchant found his way down the sides of the Cordilleras, wandered over the sunny regions at their base, and brought back the luxuries which nature had denied to his own."

In Mexico the Spaniards found an advanced civilization, but with certain weaknesses which made conquest easier. Repression was evident in the observation of "a vigilant police which repressed everything like disorders among the people." in battles, both sides employed terrain advantages. Bad weather created tactical problems: "When they had reached the opposite side, they had new impediments to encounter in traversing a road never good, now made doubly difficult by the deep mire and the tangled brushwood with which it was overrun." In defeat "the retreating army held on its way unmolested under cover of the darkness. But, as morning dawned, they beheld parties of the natives moving over the heights, or hanging at a distance, like a cloud of locusts on their rear."

In tactical maneuvering: "Three routes presented themselves to Cortes, by which he might penetrate into the valley. He chose the most difficult, traversing the bold sierra which divides the eastern plateau from the western, and so rough and precipitous, as to be scarcely practicable for the march of an army. He wisely judged that he should be less likely to experience annoyance from the enemy in this direction, as they might naturally confide in the difficulties of the ground for their protection." After leaving Chalco, the Spaniards passed through deep gorges with fortified cliffs and "the occupants of these airy pinacles took advantage of their situation to shower down stones and arrows on the troops, as they defiled through the narrow passes of the sierra." Under siege the Indians "supported life as they could, by means of such roots as they could dig from the earth, by gnawing the bark of trees, by feeding on the grass...." And "their only drink was the brackish water of the soil, saturated with the salt lake." It is little wonder that the survivors and their descendents had some unpleasant memories of the conquest.

IRAN The current war (at the time of this writing, February 1983) between Iran and Iraq has been much influenced by the soils and the hostile environmental conditions. Thousands of fighters on both sides have been killed when trapped on exposed terrain or bogged down in soils with poor trafficability. Huge amounts of funds have been wasted by military expenditures on both sides that the countries can ill afford. In 1980, the hostage situation was handled poorly by both the United States and Iran due to lack of communication and understanding (Anderson, 1980a, b; Anonymous, 1980a, b; Olson, 1980). In fact, President Carter was probably defeated in the 1980 U.S. election partly because of his demonstrated lack of understanding of the Iranian environment and the behavior patterns of the people in that environment (Anderson, 1980a, b; Olson, 1980). The U.S. airborne penetration of Iran to rescue the hostages failed primarily because of poor selection of the route and the landing site, lack of understanding of the soils and environmental conditions, and disabling of helicopters during sandstorms (a soil situation) due to inadequate air filters on the engine intakes. The timing of the hostage release (Anonymous, 1981) helped in the election of President Reagan. The Iran experience illustrates the influence that soil conditions in faraway places have even upon the U.S. political elections. Better knowledge of soils and the environment in other countries would improve foreign policy decisions in the future, and enable better prediction of human behavior in those environments.

WATERLOO The outcome of specific battles is determined largely by the soils and the environment (commonly called "terrain" by military campaign commanders) and the manner in which the different adversaries use their comparative terrain, troop, and equipment advantages. The Battle of Waterloo on June 18, 1815, provides a good illustration of the effects of soils on battles (Stewart, 1974). At Waterloo, the Duke of Wellington commanded Allied forces (from Austria, Prussia, Russia, and England) against the French forces commanded by Napoleon; the battle took nine hours and the lives of 65,000 men at a site of about 3 square miles of gently rolling farmland in Belgium. Heavy rains fell at the site on the night before the battle; as a result, the soils were saturated and trafficability was very poor, especially in some of the wetter, lower places. Some of the soil

situations contributing to the outcome of the battle have been excellently described by Stewart (1974):

> In ordinary circumstances the French counted on their cannonballs to land and ricochet on beyond the impact point, but so sodden was the earth that when the huge balls struck they sank at once in the mud. Artillery barrages were followed by cavalry charges, and here too the weather (and the soils) played against the French. The night before, many of them, rather than bed down on the rain-soaked ground, had slept atop their horses. As a result, on the day of the battle the horses were so exhausted from bearing weight all night that in the heavy going they were able to advance at little more than a trot against the Allied artillery blasts.

In the higher defensive positions the dead and dying were piled upon one another and covered the ground. One survivor that night "told how his barefoot journey back to his own lines was mostly over jellied lumps of human flesh."

Ironically, even the present large monument at the battlefield is affected by the soil conditions (Stewart, 1974):

> In 1820, the Dutch decided to raise a monument in honor of the Prince of Orange, who at the age of 23 fought and was slightly wounded in the battle. The monument that resulted is a huge four-sided earthen pyramid, 500 feet high, topped by a mammoth bronze lion....The passing decades have caused the turf to settle unevenly along the pyramid's steep sides. In an effort to stem further erosion, small trees and shrubs have been planted at varying levels. The effect is of untweezered whiskers on an old woman's chin.
>
> From the summit on a clear day it is easy to see all of the battlefield from Wellington's line on one side down to where the last of the French troops waited in reserve until the evening call to charge. The rooflines of La Haye Sainte and Hougoumont are also seen. If the day is sunny and mild, particularly on Sundays, the lush green of the acreage around the monument will be dotted with picnicking families, their napkins and newspapers reduced to tiny white flutters, the laughter of their children, the barking of their dogs, just barely audible on the sweet country air. The pleasant pastoral scene belies the carnage that took place here

on that fateful June Sunday long ago, a carnage that Wellington himself best summed up. "Nothing except a battle lost," he wrote, "can be half so melancholy as a battle won."

WORLD WAR II During World War II great efforts were expended to prepare soils information for invasions and military campaigns (Cady et al., 1945; Figures 107 and 108). Some of these efforts were continued after the war (Orvedal, 1982). Many relevant references on military applications of soils information are in the geologic and engineering literature (Coates, 1981; DOA, 1952, 1959a–c, 1967; Murdoch et al., 1971; Olson, 1973; Scott et al., 1971; Stewart, 1968). Many soils and terrain studies were also conducted during the Korean and Viet Nam conflicts (DOA, 1959a–c; Meyers, 1966; Kolb and Dornbusch, 1966; Rojana-soonthon, 1966; Wallenhorst, 1969). Much work was done on soils in relation to nuclear weapons testing and nuclear fallout (Christenson et al., 1958; DOA, 1959a, 1963; Eisenbud, 1959; Evans, 1958; James and Menzel, 1959; Mortensen, 1961; Olson, 1964; Pratt and Cooper, 1968; Schulz et al., 1959; Spitsyn and Gromov, 1959; Welford and Collins, 1960). Recently, remote sensing has been an area of military intelligence that has received considerable emphasis (Rudd, 1974).

Military campaigns are highly dependent on engineering characteristics of soils, especially in construction of roads and airfields and terrain and trafficability studies. The Unified soil classification system was developed through engineering work with many diverse soils during World War II (see pages 46–48 of the textbook). After World War II, it was revised and expanded in cooperation with the U.S. Bureau of Reclamation, so that it currently applies to embankments and foundations as well as to roads and airfields. The Department of the Army Technical Manual 5-545 (DOA, 1967), for example, includes a summary of the Unified system, with instructions to field commanders for estimating engineering soil classes rapidly when laboratory data are not available. Pedological soil classes are readily translated into Unified soil classes, as Table 67 illustrates. Basic soil texture classes and Unified classes are arranged in approximate order from best to worst for average trafficability conditions in Table 67. The "X" symbols mark the approximate equivalents in both systems.

APPLICATIONS Soil survey interpretations for military campaigns are illustrated by terrain studies conducted for training around the West Point Military Academy and other military bases. Soil maps were specially prepared on USGS topographic maps at a scale of 1:24,000. Soil trafficability was related to texture, drainage class, consistence, and so on—and to the Unified soil engineering classes. Slopes, wetness, bedrock depth, rock outcrops, stoniness, and so on, are critical soil factors for troop and vehicle movements. In fact, maps can be made for tactical movement of specified vehicles which have different capabilities (e.g., snowmobile, Jeep, tank, armored personnel carrier, amphibious vehicle, truck, motorcycle). Organic (muck) soils and other problem areas were delineated. Tide and flood levels are important, and must be specified for different times (diurnal and seasonal). Also important are wind direction, rainfall, snowfall, sandstorms, and frost. Land use must also be considered: size and density of trees is a determinant of ease of passage of different vehicles. Tanks can knock down small trees, but Jeeps cannot. Rice paddies and other fields with seeded crops damaged by maneuvers of heavy tracked vehicles are not likely to win the "hearts and minds" of the local population. Each soil also behaves differently under the impact of artillery shelling, so that explosions and damage are different for each soil map unit. Generally, the forces that have the best knowledge of the local soils and the environment and the terrain are the victors in battles and wars, if other military strengths are comparable. Of relevance is the saying that "the army with the fanciest uniforms always loses the war" (quoted from an anonymous source).

The Cornell Aeronautical Laboratory (Wallenhorst, 1969) developed techniques for making terrain trafficability maps, utilizing a computer, for an "Off-Road Mobility Research Program." Performance capabilities of five vehicles were evaluated:

1. The M-113 armored personnel carrier (tracked)
2. The M-41 five-ton, 6 X 6 utility truck (wheeled)
3. The five-ton, 4 X 4 Goer (wheeled)
4. The 2½-ton, 10 X 10 vehicle built for the Army's Mobility Exercise A (wheeled)
5. The M-29C Weasel (tracked)

For each vehicle, computer maps were made of surface composition (soils), surface geometry (topography), vegetation, and hydrology. Classes of the factors were:

1. Surface composition: 10 classes of soil strength, taking into account softness (Army Engineers rating cone index, cohesion, and shear).
2. Surface geometry: 40 classes, taking into account slope, obstacle shape, and obstacle spacing.
3. Vegetation: 18 classes, using the average diameter of the trees and the average spacing between them.
4. Hydraulic geometry: 60 classes, using 7 factors for high water and 6 for low water.

Computer maps (Wallenhorst, 1969) were divided into 40,000 squares, each 100 meters on a side. For each vehicle a "go/no go" map was constructed, showing a commander not only where his vehicles could not get through, but also why they could not. With this information a vehicle could be matched to the terrain. Common practice is to design a vehicle and then—through tests—find out what it can do. A better alternative would be to specify how the vehicle must perform in a given environment.

NUCLEAR WAR The horrors of modern weapons (including chemical, biological, radiological weapons) have such devastating effects upon human beings and the environment that new morals and ethics must be applied in future decisions of military campaigns. Much greater knowledge of soils and the environment will be needed in the future than has been employed in the past. Strategy and tactics must be considerate of long-range survival (Sidorenko, 1970). In fact, human survival may be directly dependent on knowledge of soil characteristics. A classic quotation (Scheer, 1982) is that of T. K. Jones, Deputy Under Secretary of Defense for Strategic and Theater Nuclear Forces regarding nuclear war: "Dig a hole, cover it with a couple of doors and then throw three feet of dirt [soil] on top....It's the dirt [soil] that does it....if there are enough shovels to go around everybody's going to make it."

ASSIGNMENT For your area of interest, prepare a soil trafficability map that shows where a medium tank (or other military vehicle) can move and where it cannot move within the area. Show significant soil characteristics including slopes, wetness, stoniness, rockiness, and so on. Locate an airfield and road network within the area. Locate and design a fallout shelter in case of nuclear attack. Show where you would put troop and artillery emplacements in defensive positions. Consider several hypothetical tactical situations for offensive deployment of troops and equipment. Write a brief report (including maps) about your strategic and tactical considerations of the soils and the environment in your military campaign.

As an alternative assignment, research a battle area and write a brief report on the role that soils played in the outcome. Battles of the Civil War would be good topics, because a great deal of information is available about the conflicts and soil survey reports for the areas are also readily available. Remember to consider both the long-range strategic situation and the short-range tactical maneuvers. Sherman's march to the sea, for example, involved large-scale strategy to destroy supplies and communications; the Battle of Gettysburg, in contrast, was within a relatively smaller area.

REFERENCES

Ackerman, F. et al. 1970. The Indochina story. Bantam Books, New York. 347 pages.

Anderson, J. 1980a. A man in the middle of the Iranian crisis is the man who forecasted it. The Ithaca Journal. 4 Feb. Page 5.

Anderson, J. 1980b. Invasion of Iran planned: Who is Carter trying to rescue? The Ithaca Journal. 18 Aug. Page 8.

Anonymous. 1980a. U.S. blockade means war: Iran-Vance won't rule out action. Daily News (NY). 12 Jan. Page 1.

Anonymous. 1980b. Carter aide ties defeat to hostage crisis in Iran. The New York Times. 21 Dec. Page 3.

Anonymous. 1981. Hostages coming home. The Ithaca Journal. 19 Jan. Page 1.

Anonymous. 1983a. Famine again perils the Sub-Saharan region. The New York Times. 6 Feb. Page 7.

Anonymous. 1983b. Shifting priorities. The New York Times. Page 2E.

Apple, R. W. Jr. 1982. British said to take ridge close to Stanley airfield. The New York Times. 8 June. Page A8.

Basham, A. L. 1954. The wonder that was India. Grove Press, Inc., New York. 568 pages.

Cady, J. G., M. M. Striker, and V. P. Sokoloff. 1945. Application of soil science in terrain intelligence studies. Soil Science Society of America Proceedings 10:371–374.

Chira, S. 1982. Israeli Premier visits captured castle: Crusaders built the castle. The New York Times. 8 June. Page A15.

Christenson, C. W. et al. 1958. Soil adsorption of radioactive wastes at Los Alamos. Sewage and Industrial Wastes 30:1478–1489.

Coates, D. R. 1981. Environmental geology. John Wiley & Sons, Inc., New York. 731 pages.

DOA. 1952. Geology and its military applications. Technical manual TM 5-545. Dept. of the Army, Washington, DC. About 300 pages.

DOA. 1959a. Chemical composition and neutron-induced radioactivity potential of selected soils and rocks. Engineering Intelligence Note 32, Military Geology Branch, U.S. Geological Survey, Washington, DC. 235 pages.

DOA. 1959b. Terrain study of Fort Benning and vicinity. Engineering Intelligence Study 211, Chief of Engineers, Dept. of the Army, Military Geology Branch, U.S. Geological Survey and Soil Survey, U.S. Dept. of Agriculture, Reston, VA. 79 pages (folio).

DOA. 1959c. Production of cross-country movement studies. Engineering Intelligence Guide 31, Military Geology Branch, U.S. Geological Survey and Soil Conservation Service, U.S. Dept. of Agriculture, Washington, DC. About 100 pages.

DOA. 1963. General guide for estimating significant soil characteristics for predicting gamma hazard from neutron-induced activity. Prepared for the Defense Intelligence Agency, U.S. Dept. of Defense by World Soil Geography Unit, Soil Survey, Soil Conservation Service, U.S. Dept. of Agriculture, Washington, DC. 164 pages and maps.

DOA. 1967. Geology. Technical Manual 5-545. Dept. of the Army, Washington, DC. About 50 pages.

Eisenbud, M. 1959. Deposition of Strontium 90 through October 1958. Science 130:76–80.

Eisenhower, D. D. 1948. Crusade in Europe. Doubleday & Company, Inc., Garden City, NY. 559 pages.

Evans, E. J. 1958. Chemical investigations of movement of fission products in soils. Chalk River Project No. 667. Atomic Energy Canada Ltd. (Also in Chemical Abstracts 53:2840.)

Fisher, W. B. 1963. The Middle East: A physical, social, and regional geography. Methuen, London. 568 pages.

Foth, H. D. and J. W. Schafer. 1980. Soil geography and land use. John Wiley & Sons, Inc., New York. 484 pages.

Gettleman, M. E. (Editor). 1965. Viet Nam: History, documents, and opinions on a major world crisis. A Fawcett Crest Book, Greenwich, Conn. 448 pages.

Gunther, J. 1955. Inside Africa. Harper & Brothers, New York. 952 pages.

Hambly, G. 1969. Central Asia. Delacorte Press, New York. 388 pages.

Hanke, L. 1968. Contemporary Latin America: A short history. D. Van Nostrand Company, Inc., Princeton, NJ. 532 pages.

Hazarika, S. 1983. Violence seems to grow in India's troubled state. The New York Times. 18 Feb. Page A3.

Inman, B. R. 1982. Transcript of statements at State Dept. of the military buildup in Nicaragua. 10 Mar. Page A16.

James, P. R. and R. G. Menzel. 1959. Radioactive fallout: How soils are decontaminated. Plant Food Review 5:5–7, 32–33.

Katz, F. (Coordinator). 1983. A matter of survival: Whoever controls the strategic Judea and Samaria controls Israel. The New York Times. 14 Feb. Page A15.

King, S. S. 1982. Food sales by U.S. called peace tool: Report regards the Russian's reliance on American grain as a nuclear deterrent. The New York Times. 3 Oct. Page 11.

King, J. C. and J. A. McGilvray. 1973. Political and social philosophy: Traditional and contemporary readings. McGraw-Hill Book Company, New York. 563 pages.

Kolb, C. R. and W. K. Dornbusch, Jr. 1966. Analogs of Yuma terrain in the Middle East desert. Technical Report 3-630, U.S. Army Engineer Waterways Experiment Station, Vicksburg, MS. Volume 1 (52 pages). Volume 2 (Map folio of 19 plots 23″ X 30″).

Metzger, B. M. 1982. The Reader's Digest Bible. The Reader's Digest Association, Pleasantville, NY. 799 pages.

Meyers, M. P. 1966. Comparison of engineering properties of selected temperate and tropical surface soils. Technical Report 3-732, U.S. Army Engineer Waterways Experiment Station, Vicksburg, MS. 234 pages.

McNeill, W. H. 1983. The pursuit of power: Technology, armed force, and society since A.D. 1000. University of Chicago Press, Chicago. 405 pages.

Michener, J. A. 1951. The voice of Asia. Random House, New York. 245 pages.

Middleton, D. 1982a. Forty years ago, the battle for North Africa. The New York Times Magazine. 7 Nov. Pages 48–68.

Middleton, D. 1982b. British victory: Coordination and professionalism. The New York Times. 16 June. Page A23.

Moreland, W. H. and A. C. Chatterjee. 1962. A short history of India. David McKay Company, Inc., New York. 594 pages.

Mortensen, J. L. 1961. Radioactive fallout. Agronomy Journal 53:343-348.

Murdoch, G., R. Webster, and C. J. Lawrance. 1971. Atlas of the land systems of Swaziland. Military Vehicles Eng. Establishment, Barrack Road, Christchurch, Hants, England. 49 pages in loose-leaf folder.

Nguyen Hoai Van. 1962. Inventory and nomenclature of Viet Nam soils. M.S. thesis, Dept. of Agronomy, Cornell University, Ithaca, NY. 134 pages and soils maps.

Olson, G. W. 1964. Application of soil survey to problems of health, sanitation, and engineering. Memoir 387, New York State College of Agriculture, Cornell University, Ithaca, NY. 77 pages.

Olson, G. W. 1973. Soil survey interpretation for engineering purposes. Soils Bulletin 19, Food and Agriculture Organization of the United Nations, Rome, Italy. 24 pages.

Olson, G. W. 1980. The strategic situation in Iran according to the soils. Soil Survey Horizons 21:21–24.

Orvedal, A. C. 1982. Personal communication dated 28 Jan. Several pages in letter and enclosures.

Owens, E. and R. Shaw. 1972. Development reconsidered. D. C. Heath and Company, Lexington, MA. 190 pages.

Pratt, H. R. and H. R. Cooper. 1968. The near surface geology of Eniwetok and Bikini atolls. Report AFWL-TR-68. Weather Lab, U.S. Air Force, Kirkland Air Force Base. 84 pages.

Prescott, W. H. 1934. Conquest of Mexico. The Book League of America, New York. 488 pages.

Rojanasoonthon, S. 1966. Great soil group survey for selected study areas in Thailand—Volume 1 (Summary Report). Volume 2 for various study areas. Sponsored by Advanced Research Projects Agency, Directorate of Remote Area Conflict, U.S. Army Material Command. Coordinated for U.S. Army Engineer Waterways Experiment Station, Corps of Engineers, Vicksburg, MS. By Kasetsart University, Bangkok, Thailand. Several hundred pages.

Rudd, R. D. 1974. Remote sensing: A better view. Duxbury Press, North Scituate, MA. 135 pages.

Scheer, R. 1982. With enough shovels: Reagan, Bush, and nuclear war. Random House, New York. About 300 pages.

Schmidt, B. L. (Chairman) et al. 1982. Determinants of soil loss tolerance. Special Publication 45, American Society of Agronomy, Madison, WI. 153 pages.

Schulz, R. K. et al. 1959. Some experiments on the decontamination of soils containing Strontium 90. Hilgardia 28:457–475.

Scott, R. M., R. Webster, and C. J. Lawrance. 1971. Atlas of the land systems of western Kenya. Military Vehicles Engineering Establishment, Barrack Road, Christchurch, Hants, England. 363 pages in loose-leaf folder.

Sidorenko, A. A. 1970. The offensive (a Soviet view). Translated and published under the auspices of the U.S. Air Force, U.S. Government Printing Office, Washington, DC. 228 pages.

Spitsyn, V. I., and V. V. Gromov. 1959. Adsorption of radiostrontium by some minerals of soils and grounds. (In Russian.) Pochvovedenie 12: 45-50. (Abstract in Soils and Fertilizers 23:494).

Stamp, L. D. 1965. A history of land use in arid regions. United Nations Educational, Scientific, and Cultural Organization, Paris, France. 388 pages.

Stewart, G. A. (Editor). 1968. Land evaluation: Papers of a CSIRO symposium organized in cooperation with UNESCO 26–31 Aug. Macmillan of Australia, Sydney. 392 pages.

Stewart, L. 1974. The Battle of Waterloo: Sunday picnics on the killing ground. The New York Times (Travel Section). 23 June. Pages 1 and 22.

Thomas, W. L. Jr. et al. 1965 (5th impression). Man's role in changing the face of the earth. University of Chicago Press, Chicago. 1193 pages.

Times. 1982. The Times concise atlas of world history. Hammond, Inc., Maplewood, NJ. 192 pages.

Von Hagen, V. W. 1964 (4th printing). World of the Maya. Mentor Books, New York. 222 pages.

Wallenhorst, R. (Editor). 1969. Improving off-road mobility. Research Trends 17:3-8. Cornell Aeronautical Lab, Buffalo, New York.

Welford, G. A. and W. R. Collins, Jr. 1960. Fallout in New York City during 1958. Science 131:1711–1715.

FIGURE 106/*Soldiers in Italy pushing their Jeep out of the mud (adapted from Eisenhower, 1948).*

FIGURE 107/*Hill on Okinawa heavily fortified by the Japanese in caves in World War II. The hill has soils shallow to weakly consolidated calcareous shales. The valley in the foreground has clayey alluvial soils with poor trafficability characteristics when wet. The many American casualties suffered in the military campaign to capture this hill would have been fewer if better knowledge of soil characteristics had been employed (photo by Roy Simonson).*

FIGURE 108/*Clayey alluvial soils on Okinawa with poor trafficability characteristics for military campaigns. Heavy vehicles became bogged down in these wet, sticky clay soils when American forces tried to capture the island from the Japanese occupiers. The remains of a destroyed American tank can be seen to the right of the upper part of the muddy trail. Better knowledge of the soil conditions might have saved the tank and the crew during the military campaign (photo by Roy Simonson).*

TABLE 67/General trafficability materials comparison of Unified engineering and pedological soil texture classes. Approximate equivalents are marked with "X" symbols. (Adapted from DOA, 1959; Orvedal, 1982.) See pages 22 and 46 in the textbook. The "XX" symbols indicate an equivalent class only when the soil map unit is gravelly, cobbly, or stony and the coarse aggregate of such soil is greater than 50 percent. Soil drainage and characteristics other than texture are not considered in this table.

U.S. Army Corps of Engineers Unified System			U.S. Department of Agriculture Soil Texture Classification											
Trafficability Classification	Unified Symbol	Description (abbreviated)	Sand S	Clay C	Silty clay SiC	Silty clay loam SiCL	Clay loam CL	Sandy clay SC	Sandy clay loam SCL	Sandy loam SL	Loamy sand LS	Loam L	Silt loam SiL	Silt Si
Good														
A	GW	Well-graded gravel, sand mixture												
A	GP	Poorly graded gravel, sand mixture												
A	SW	Well-graded sand, gravel mixture	X											
A	SP	Poorly graded sand, gravel mixture	X											
Fair														
B	CH	Inorganic clays, fat clays		X	X	X	X							
Poor														
C	GC	Clayey gravel, mixture						XX	XX	XX				
C	SC	Clayey sand, mixture						X	X	X				
C	CL	Inorganic clay, mixture				X	X	X	X					
Very poor														
D	GM	Silty gravel, mixture	XX							XX	XX			
D	SM	Silty sand, mixture	X							X	X			
D	ML (ML-CL)	Inorganic silty and very fine sand										X	X	X
D	MH	Inorganic silts, etc.		X									X	X
D	OL	Organic silts, low plasticity										X	X	X
D	OH	Organic clays, high plasticity		X	X	X	X							

Research

This Field Guide and the accompanying textbook can be used also as a guide to further research on soils and the environment. Many of the exercises, projects, and topics discussed can be expanded or narrowed into suitable subjects for M.S. and Ph.D. theses subjects. These Field Guide exercises were designed to acquaint students, teachers, and laypersons with a broad range of possibilities of soil survey interpretations, and no one student, person, or class could be expected to exploit fully all of the exercises in the amount of time available to most people. Users of this Field Guide are encouraged to concentrate on the topics of interest and greatest relevance to them. Probably the greatest opportunities for research in the future are in the areas of interdisciplinary linkages of data from other fields to soil map units. Thus, many people can benefit from correlating their data to soil map units, including agriculturalists, agronomists, assessors, botanists, conservationists, contractors, ecologists, economists, engineers, extension workers, foresters, geologists, groundwater experts, planners, politicians, public health officials, range managers, recreationists, wildlife specialists, and many others. Research can be formal or informal: it can be structured toward an M.S. or Ph.D. degree, or it can be merely an individual's relating of yields or some other data measurements to soil map units of the soil survey. Everyone can be involved in research: research is simply the gathering of data and organization of the information into a format that did not exist previously. Research is defined as critical and exhaustive investigation or experimentation having for its aim the revision of accepted conclusions, in the light of newly discovered facts.

Part I of this Field Guide deals with introduction to the language of soil survey interpretations and the criteria by which soils are rated for different purposes. This part is basic, and a great deal of research is needed to improve further soil profile descriptions, soil maps, legends, lab analyses, computerized groupings of soils, and simple ratings of the information for practical use. Soil profile descriptions can always be improved, and new techniques are needed to better describe ranges in soil characteristics within soil map units and measurements of properties of soils not commonly observed (e.g., bearing strength, Pb, Zn, dioxin, yields, land-use correlations). Better expressions of effects of scale on reliability of soil maps are needed, so that people can more adequately relate data of general soil maps at small scales (e.g., of states and regions) with highly detailed soil maps at large scales (e.g., for experiment stations and irrigation project areas). Many research opportunities exist for formulating improved computerized groupings of soils. Soil Taxonomy and soil survey interpretations must be used more widely in computer programs in the future, especially in data storage and retrieval and in map display. Computer correlations of soil performance data (e.g., yields, foundation breakup, etc.) will be especially valuable research topics for the future. Practical projects always involve considerable research if they are done well in a comprehensive format. Students investigating an area for a project should have a natural curiosity to find out all they can about the area that can be related to soil survey interpretations.

Part II of this Field Guide involves applications of soil surveys in systems of wide usage. Testing of these systems, of course, is in itself a form of research. When students place soils of their project area into land capability units, or calculate soil loss from a specific map unit—that is a type of soil research, because that information did not exist before in that format. Yield measurements and correlations to soils are research. Engineering applications only begin with routine soil interpretations, and then lead into deeper borings and interpretations of elaborate engineering soil tests. Waste disposal in soils must consider effluent and pollutant movements, and long-term monitorings of toxic materials in soils and landscapes, if human health and the environment are to be protected. Land classification systems need to be continually revised and improved, and that improvement is research to gather new data that did not exist before to improve the systems. The entire study of erosion control is very complex, and should involve anthropologists, economists, sociologists, engineers, geologists, politicians, and many others as well as soil scientists doing research on soils and yields.

Using soil surveys for taxation requires many more data than now exist if taxation systems are to be kept current and equitable. Computer modeling of watersheds and drainage basins must be expanded in the future, if we are to understand landscape processes and maintain and improve the soils and the environment. Farm planning and community planning must not be merely "cookbook" procedures, but must involve research specific to the site and to the development to ensure the correct decisions for achieving maximum efficiency for landscape management in the future.

Part III of this Field Guide illustrates principles governing the applications of soil survey interpretations in the future. Most of these exercises and topics deal with data gathering for research in soil survey interpretations. The economics and feasibility of corrective measures must be incorporated into evaluations of future "soil potentials." Soil variability, of course, is a vital topic that is extremely important for understanding performance of soil map units, and must be researched and characterized statistically so that probability statements can be quantified for specific soils. Soil variability characterization and description is of extreme importance in interpreting yield results from experiment stations, and in "technology transfer" in extension of data to other areas. The transfer of research data through soil correlations will enable yield comparisons with other areas of similar soils, and contrasts with areas of dissimilar soils. Sequential testing in soil research will enable greater efficiency in identifying yield differences due to soil characteristics and erosion, and in isolating variables due to climate (weather) and management. Computer correlations of land uses and soils in research projects will enable a better understanding of soil effects over long periods of time, and enable probability predictions to be made about land-use shifts in the future.

Long-term research on land use and erosion will help to identify situations that may revert to environmental degradation in the "tragedy of the commons," and suggest remedies whereby collapse of ecological and environmental systems can be avoided. With comprehensive research perspectives, strategic implications will be clarified in planning for the future. Hopefully, soil information will not be needed for military campaigns, but we should be ready with soil interpretations for tactical considerations just in case. Greater understanding of soils and environments will help us to understand better behaviors of different peoples in other countries, and will assist in foreign policy decisions. We can predict the future, based on knowledge of our resources and their best uses. If soils and other resources are used properly (physically, chemically, morally, ethically), a society based on them will probably persist and prosper; if soils are excessively exploited and abused, that society based on that abused resource cannot survive. Of prime importance in the wise use of soil resources is the education of everyone about their soils, so that public and private decisions can be made for the optimum benefits of the individuals and for society as a whole.

Predictions

The chief value of information on soils and the environment is in the predictions that can be made from the data. Soil descriptions, maps, lab data, and classifications all enable the environment to be characterized so that we know what will happen under stress or under enhancement. If a forested area is clearcut, for example, we know that the soils will erode differently in various places—and we can predict the soil loss from each area with considerable precision. Conversely, we know that mulching, conservation tillage, and cover crops will reduce erosion until trees can be replanted. Using soil maps, we can determine the best routes for roads to avoid excessive grades and gullying. Using soil maps, we can plan farms and communities so that maintenance and operation will be most efficient, in all aspects involving spatial (areal) considerations. We can avoid land-use problems, correct problems observed in early stages, and have a more aesthetic and efficient human environment with forethought and consideration of our soil resources.

With soil maps, we can design foundations and locate gravel sources. We can plan waste disposal in lagoons or landfill with minimal hazards. Landscape segments (soil map units) can be designated for equitable land taxation. Soil loss through erosion can be predicted, and control measures can be established where losses are excessive. Yields (of crops, range, forests, etc.) can be correlated with specific soil map units for long-term considerations. Field boundaries on farms can be shifted, to enlarge cropping patterns in accord with natural landscape contours—and terraces, dams, waterways, drainage systems, and other structures can be located in harmony with the environment. Community parks, playgrounds, streets, housing developments, industries, shopping centers, and agricultural sustaining areas can be planned and managed most efficiently in accord with the soils and the environment.

With research, we can improve our understanding of soil variability through sequential testing and land-use corrrelations. Predictions can be increasingly quantified into useful probability statements of scenarios relating to specific soil map units. With increased understanding of soils and the environment, we will be able to predict strategic implications in soil use and abuse important for national security and relations between and within nations. The concept of the "tragedy of the commons" enables us to predict the collapse of environmental systems, and prevent the collapse if we would only maintain and improve our environment. In military campaigns, we can predict tactical actions and influence the outcomes of battles: offensive tank attacks are likely to fail across wet, clayey plains in rainy seasons; occupation forces imposing excessive land taxes irrespective of soil quality are not likely to win the "hearts and minds" of the local population; dictatorships that appropriate the best soils for estates of a few rich people while most of the population subsists poorly on marginal soils are likely to prove to be unstable forms of government; and countries that have good soil resources and good management programs (and good moral judgments) for long-range, large-scale development and maintenance of their soils and the environment are likely to be stable and persist for a long time. Hopefully, this textbook and Field Guide will assist in helping people to think about their soils and their environment, and place the high priority on the improvement and management of the soils that the resources deserve.

Soils tours

Soils tours provide the best mechanism to teach students, special interest groups, and the general public about soils. During a tour of soils in the field, each participant can relate the soil map and report to soil profile and landscape characteristics according to each individual's perspective. In New York State, a systematic sustained program of soils tours is under way to communicate soil survey information to decision makers. The base for the program is a series of published "Soil Tour Fact Sheets" which outline a route and observations on a soil map sheet. Published photographs and descriptions of soil conditions emphasize the important things to be observed on the tour. With the Soils Tour Fact Sheet, a county road map, the soil survey report, a shovel, and a Cornell pH test kit, a tour group can then determine the actual field conditions of the soils in relation to their location on the map. In New York State, Soils Tour Fact Sheets are under preparation in a number of counties, and have already been published for the following counties: Broome, Cayuga, Chemung, Cortland, Genesee, Herkimer, Lewis, Monroe, Montgomery, Niagara, Onondaga, Ontario, Orleans, Schenectady, Schoharie, Schuyler, Seneca, Suffolk, Tioga, Tompkins, Ulster, Washington, Wayne, Westchester, Wyoming, and Yates. A walking soils tour has even been prepared for Central Park in New York City! Following this brief description is an example of one of the published soils tours, titled "Soils Tour near Chenango Bridge, Broome County [New York]."

Through experience, the best format for the soils tour program appears to be an indoor session followed by an outdoor session on the same day. People from Cooperative Extension, the Soil Conservation Service, the Agricultural Experiment Station, and other agencies and institutions work closely together in planning each tour. Each tour group is different, and the publicity for the tour can be targeted toward a specific or a general audience. Indoors, people involved with making and using soil surveys explain how a soil survey is made, how it can be used, and specific landscape and soil character in the area to be visited. Examples of maps, aerial photographs, and soil profile monoliths are displayed in the meeting room. After lunch or a coffee break period, participants board buses or cars stopping at several locations in the tour area. At each stop the soil landscape is described in detail in relation to the views observed, the pattern displayed on the aerial photograph of the soil map, and the soil properties dug into in the roadcuts and other excavations.

Many questions are answered in this informal setting, ranging from questions about pH readings to advanced laboratory techniques and remote sensing applications. When the tour is over, everyone has been mentally and physically enlightened and challenged—and has a sense of accomplishment in learning about the surface and subsoil characteristics of the local environment. Typical evaluation comments include: "I learned more about soils today than in all my courses at Cornell!"; "I never realized we had so much clay in _____ County."; "Now I know where to look for gravel pits."; and "If only we had this information when we were planning the subdivision last year." Enthusiasm for the tours is high, and plans are to develop them and implement them in all the counties in New York State. Other states and countries could similarly develop systematic soils tour programs adapted to the needs of each community. A great deal of time and effort is involved in planning and conducting each tour, but no more effective techniques can be found than to bring the people to the object of study—*the soils.*

SOILS

COOPERATIVE EXTENSION • NEW YORK STATE • CORNELL UNIVERSITY

Page: 13.00
Date: 4-1980

Tour of Soils near Chenango Bridge in Broome County, New York

By Gerald W. Olson
Dept. of Agronomy
Cornell University

Field examinations of soils and soil maps are the best techniques to learn and teach about our most valuable resources. Each soil occupies its characteristic position in the landscape, and the variable soils determine to a large extent the present land use and the future potentials for improvements. For a tour of the soils in a county, a good procedure is to select a route with typical soil and land use conditions in a variety of landscapes, within reasonable driving distance from a place easily located. With a county road map, the soil map, a shovel, and a Cornell pH test kit, a single person or group of people can then determine the actual field conditions of the soils in relation to their location on the map. Soils are mapped through the Cooperative Soil Survey—a joint effort of the United States Department of Agriculture, the Agricultural Experiment Station, and other agencies in each state. In New York State, workers of the Soil Conservation Service, Cooperative Extension, Cornell University, and other agencies and institutions are usually available to help people with soils tours. In the following description, map, and photos, a tour of the soils is outlined for an area around Chenango Bridge in Broome County. This tour is one example of many tours that can be used in different counties to illustrate the soil conditions typical of the variable landscapes of New York State.

The area around Chenango Bridge illustrates excellently some of the major contrasting soil conditions between upland soils and valley soils in Broome County. The urban and rural development conforms (on a statistical probability basis) to the soil conditions, and soil performance over the years has dictated success or failure of farming operations, septic

Figure 1. *ChD* Chenango and Howard gravel bank at Northrup Industries near Chenango Bridge. These soils are excellent for most uses and have been graded or filled at many locations in Broome County. The surface soil has not been graded here because a small cemetery occupies the top of the glacial terrace.

Figure 2. Industrial and commercial development on *ChA* Chenango and Howard soils in Hillcrest. These soils are nearly level, well drained, and have excellent bearing capacity. They also absorb sewage effluent rapidly and have good groundwater aquifers.

tank performance, road breakup, and many other events influenced by the soil and land characteristics.

1. Follow Route 81 north from the Farm and Home Center at 840 Front Street and take the first exit to Chenango Bridge. Consult soil map sheet 42 of the Broome County Soil Survey. Follow Route 11 and 12 through Chenango Bridge, but stop before the railroad bridge to dig in and look at the *Ta* Tioga alluvial soils and the *ChD* Chenango and Howard gravel bank at Northrup Industries. Check the pH in these soils with the Cornell pH test kit. The Tioga alluvium is somewhat higher in pH than the surrounding uplands, because of alluvial (flood deposited) enrichments from losses from erosion of the nearby hills and of the limestone substrata gravel carried down the valley from glacial outwash. Howard gravelly soils are calcareous in the deep subsoils in places and even cemented in some spots because of deep leachings of calcium carbonate, even when the upper part of the soil profiles are acid.

2. Notice the elevation of silty alluvial and gravelly glacial outwash terraces and the related urban development. At Hillcrest, turn east past Chenango Valley Cemetery on Nowlan Road. Note the abrupt change from the fertile soils of the valleys to the acid glacial till soils of the uplands. The land use pattern of the hills is small fields, small farms, and houses built by people who work in Binghamton and the surrounding urban areas. The *Mh* Mardin and *Vo* Volusia soils should be examined in detail with a shovel in the subsoils exposed in the roadcuts and in the surface horizons in the fields. These soils are acid and have a fragipan that perches the water tables and causes the groundwater to seep laterally across the landscapes. The mottles (spots of different colors in the subsoils) are a good indicator of the perched seasonal water tables.

3. Note the *Ad* alluvial land characteristic of the narrow valleys nestled in the hills. The *AcA* Alden and Chippewa spots are very wet and excellent for ponds. Contrast these soils with the soils of the big broad valleys just previously examined.

Figure 3. Chenango and Howard *ChA* area in valley of Choconut Creek. These soils are valuable for farming in the small valleys where urban pressures are not so great as in the larger valleys around Binghamton. These soils are very responsive to good management for crop production.

Figure 4. House on soils developed in glacial till. The subsoils are compact and have some seepage problems, but good landscape design can provide for adequate drainage and improved septic tank seepage fields.

Figure 5. Houses on Volusia soils often have beautiful views, and landscape design can improve the problems of the dense subsoil character. Well yields are usually low in glacial till and shale bedrock areas at the higher elevations in Broome County.

Figure 6. Volusia and Mardin soils on side slopes of hills in Broome County can provide good pasture. The soils are acid, and a good liming and fertilizer program increases yields considerably.

Figure 7. Many wet Volusia soils in Broome County have excellent sites for ponds and small lakes. Subsoils in compact glacial till are relatively impermeable, so that basins for water storage can be easily constructed.

Figure 8. Rural road on *LoE* Lordstown soils on steep slopes. Sandstone, siltstone, and shale bedrock is visible in the roadcuts and ditches. Trees grow well on these soils, but erosion is a problem if the vegetation is removed.

Follow Nowlan Road to the top of the hill, examining the *Mh* and *Vo* soils in the exposed pits and the roadcuts—at the top of the hill these soils are the most acid. Nowlan Road continues onto the area shown on soil map sheet 43; note the *LoE* forested area with sandstone and shale bedrock outcroppings in the ditches cut on these steep slopes. Where the land levels out, the farms appear a bit more prosperous with somewhat larger fields and numerous houses. Some of the farms are for sale and are probably not economical agricultural units—most of the farmers probably work at least part-time in the city. Follow Nowlan Road down into the valley to Route 7; then turn around and retrace the route observing the same land and soil features from the east perspective as well as the west perspective.

4. At Hinmans Corner (see soil map sheet 42) turn west on Fuller Road. Note the dense urban housing developments on *Tg* Tioga soils. Note also the *MhE* steep soils in the gorge of the drainageway; this area has great scenic potential for houses facing the steep slopes of the gorge. Check the high content of organic matter in the roadcut in the *Wd* Wayland wet soil area shown on soil map sheet 41. This *Wd* Wayland area could be excellent for development of a small lake or ponds.

5. Turn right (north) a short distance on Wilson Hill Road shown on soil map sheet 41. Note the row of houses with a beautiful view! Turn left (west) on Aitchison Road and check the pH by digging in the *VoC* and *MhC* soils at the top of the hill in the roadcut. These subsoils are acid. Many houses have been built in recent years along this road.

6. Turn south on Dimock Hill Road (soil map sheet 41) and check the *Ms* Middlebury and other soils along the route south. Note the competing land uses for soils in the medium-sized valley of Choconut Creek; some agriculture remains, but lots of houses, too, compete with agriculture for land use in this valley. From Choconut Center, turn west for views of more extensive agriculture and home lots on *VoC* soils, dam and lake developments, and forested areas,

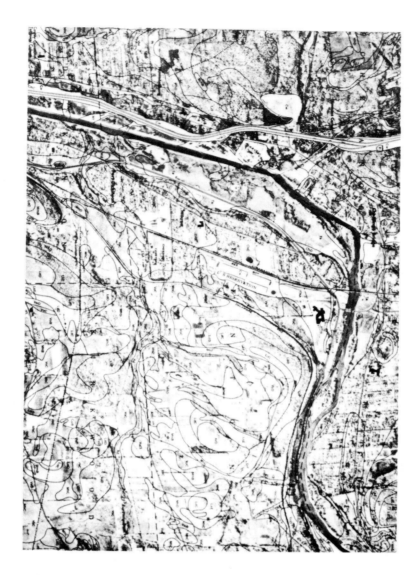

illustrating potential for forestry products development. Return to the Farm and Home Center by retracing this part of the route or by continuing south to Route 17.

References

Cline, M. G., and Marshall, R. L. 1977. *Soils of New York landscapes.* Information Bul. 119. N.Y.S. Coll. of Agri. and Life Sci., Cornell Univ., Ithaca, N.Y. 62 pages and map.

Olson, G. W. 1977. *Using soils as ecological resources.* Information Bul. 6. N.Y.S. Coll. of Agri. and Life Sci., Cornell Univ., Ithaca, N.Y. 16 pages.

Giddings, E. B.; Flora, D. F.; and Olson, G. W. 1971. *Soil survey of Broome County, New York.* Soil Conserv. Serv., USDA, in cooperation with Cornell Univ. Agri. Exper. Sta. U.S. Gov. Print. Off., Washington, D.C. 95 pages and 89 soil map sheets.

Price per copy 15 cents. Quantity discount available.

Cooperative Extension, the New York State College of Human Ecology, and the New York State College of Agriculture and Life Sciences, at Cornell University, Ithaca, N.Y., and the U.S. Department of Agriculture, cooperating. In furtherance of Acts of Congress May 8, June 30, 1914, and providing equal opportunities in employment and programs. Lucinda A. Noble, Director.   4/80 CP 2M 6781-A

Slide sets

Many slide sets are available or under preparation to assist in teaching various aspects of a course on soils and the environment. The best slide sets are also accompanied by published bulletins which show the photographs and the captions and explanations for them. Following are examples of some of the slide sets that are available:

Extension Service. 1969. Know your land. Narrative guide for color slide set and film strip, on the Land Capability Classification system. 12 pages and 50 photos. Available from Photography Division, Office of Information, U.S. Dept. of Agriculture, Washington, DC 20250.

Gonzales, N. et al. 1980. If the land dies....Color filmstrip and printed commentary, on worldwide soil erosion and its control. 70 pages and 259 photos. Available from Distribution and Sales Section, Food and Agriculture Organization of the United Nations, Via Delle Terme Di Caracalla, 00100 Rome, Italy.

Olson, G. W. 1977. Using soils as ecological resources. Information Bulletin 6, New York State College of Agriculture and Life Sciences, Cornell University, Ithaca, NY. 16 pages and 50 photos (slides). Available from Mailing Room, Research Park, Cornell University, Ithaca, NY 14853.

Olson, G. W. 1977. Using soil resources for development in Latin America. Cornell International Agriculture Bulletin 31, New York State College of Agriculture and Life Sciences, Cornell University, Ithaca, NY. 32 pages and 50 photos (slides). Available from Mailing Room, Research Park, Cornell University, Ithaca, NY 14853.

Final exam

The final exam is a "learning experience" covering all the content of the textbook and Field Guide. The format is flexible, and should be modified by each teacher to fit different classes. The primary goal for the final exam should be to encourage each student to think about application of the comprehensive content of the course. Before the exam, students should be informed about the nature of each question so that they can be well prepared for a comprehensive test. The typed project report should be due from each student at the time of the final exam, and if the student wishes to use the familiar project area for question 5, he or she should be able to do so.

Question 1 merely requests the student's address, so that the exam and course grade can be returned promptly by mail. Question 2 asks about future use of the course content from the student's personal perspective. Question 3 requests each to define the word "soil," in relation to course content and experiences in this course. Question 4 is a matching exercise that is the most "tricky" and quantitative of these questions. Emphasize that "the most related or definitive statements" in the right column are to be matched with the numbered items in the left column. Each teacher will want to modify the items in each exam, of course. The last numbers in the right column require good knowledge of the material to answer correctly; generally, students do surprisingly well on these items.

Question 5 is a comprehensive community planning exercise that can probably best be done on a plastic transparent overlay sheet that is frosted on one side—so that each student can overlay it on a soil map and write on it. Colored pencils usually work best to make the plan most legible for the teacher to grade. Students can each use the individual soil map for the student's individual project area, or the entire class (all students) can each make a community plan for the same soil map sheet (but each needs an individual copy for the exam). The exam should be a "closed book" exam, and the teacher may need to provide a legend for the map and a brief description of the soils for question 5. This last question encourages a display of initiative and imagination on the part of each student, and requires extensive thinking in application of a comprehensive knowledge of the course content.

Name _____

FINAL EXAM

(Be as comprehensive as you can in answering these questions.)

1. Your grades and project report evaluation will be sent to you as soon as possible, probably within a week or two. Please print below IN BLOCK LETTERS, THE EXACT ADDRESS to which you want these materials sent:

_____

_____

_____

_____

_____

_____

2. How do you expect to use the information you got from the course?

_____

_____

_____

_____

_____

_____

3. Define the word "soil" according to your interpretation of the definition in the materials and experiences provided to you in this course.

_____

_____

_____

_____

_____

_____

4. Place the number of the term in the left column in the blank associated with the most related or definitive statement in the right column.

1. Available H$_2$O	_____ Alfisol, Vertisol
2. Benchmark project	_____ A = RKLSCP
3. Clear	_____ Atterberg limits
4. Cobbly	_____ Cohesion, adhesion, resistance
5. COLE	_____ Drainage classes
6. Common	_____ Fine, coarse
7. Consistence	_____ Friable, firm
8. Cresol red	_____ Hohokam
9. CRIES	_____ H$_2$O fluctuation
10. Dry	_____ Inches H$_2$O/rooting depth
11. Erosion	_____ Landslides
12. Form 5	_____ lb/acre
13. Gradual	_____ Length
14. Inferred	_____ Lithologic discontinuity
15. Interpretive map	_____ % H$_2$O
16. Medium acid	_____ Regional modeling
17. Medium sand	_____ Ratings, limitations
18. Moist	_____ Shifting cultivation
19. Phoenix	_____ Slight, moderate, severe
20. Plasticity index	_____ Soft, hard
21. Platy	_____ Technology transfer
22. Quick test	_____ Thin, thick
23. Sardis	_____ Tragedy of the commons
24. Sequential testing	_____ 0.5–0.25
25. Size	_____ 1.0–2.5
26. Soil loss	_____ 2.0–20.0
27. Soil Taxonomy	_____ 2.5–5.0
28. Tikal	_____ 3.0–10.0
29. Unified system	_____ 5.6–6.0
30. II B'2x	_____ 7.2–8.8

5. Select an area (on a single soil map sheet) in your county which is likely to have many land-use changes in the next few years. On a frosted transparent overlay, plan an additional hypothetical community of about 1,000 acres for about 5,000 people. Assume adequate financing, and base your decisions mainly on the soil characteristics of the area from the soil survey. Neatness and clarity will be major considerations in grading. If necessary, state assumptions or reasons for decisions. Items of construction and land use to be located are listed below:

Assumptions: 1. _____

2. _____

3. _____

4. _____

Constructions to be located:

Residential: Mobile home park
Large lot development with septic tanks
Cluster developments
Townhouses
High-rise apartments
Sanitary landfill sites

Commercial: Shopping center
Education: Elementary school
Recreation: Small golf course
Three small lakes
One short ski slope
Picnic areas
Overlooks
Nature study area
Trailer campsite

Industry: Small electronics plant
Gravel pit
Public Buildings: Library
Firehall
Police station
Water treatment plant
Sewage treatment plant

Transportation: Main access road
Feeder roads

Farmland: Delineate 1,000 acres of the best agricultural soils for sustained farm production in an agricultural preserve (Agricultural District)

Evaluation

The following pages have a format for "Student Appraisal of Courses and Teachers" that has been extensively used in the College of Agriculture and Life Sciences at Cornell University. The first seven questions deal with background information (some questions may need modification at different institutions), and questions 8 to 18 are a general student evaluation of the teacher and the course. In addition, teachers can add questions to evaluate readings, papers and projects, fieldwork, exams, and so on. Examples of some additional questions are given in items 19 to 46. Useful also are questions where students can comment freely in writing about parts of the course. A suggested format is to provide space for student writing under several topics including:

I. What recommendations do you have about the lectures? (e.g., content, organization, delivery, guest lectures)
II. The readings; (e.g., content, scope, interest value)
III. The papers and project (e.g., content, learning experience)
IV. Laboratories? Fieldwork? Discussions? (e.g., content, value to your education)
V. The examinations? (e.g., content, style, grading, feedback)
VI. Summary
   A. The best part of this course was:
   B. The worst part of this course was:
   C. The most important change to make is:

Course name and number _____ Date _____

College of Agriculture and Life Sciences

Student Appraisal of Courses and Teachers

You are asked to respond to the following questions to provide the college with one measure of the success of this course. Your constructive criticism is appreciated. Your response will be used, in part, to make administrative decisions regarding your instructor.

Background Information

DO *NOT* FILL IN YOUR NAME OR I.D. NUMBER. THIS QUESTIONNAIRE SHOULD REMAIN ANONYMOUS. Place the appropriate answer code number in the box adjacent to each question number.

☐ 1. Sex (optional): 1 = Male   2 = Female

☐ 2. School: 1 = Agriculture & Life Sciences   6 = Human Ecology
                2 = Architecture   7 = Industrial & Labor Relations
                3 = Arts & Sciences   8 = Unclassified or other (e.g., extramural, faculty,
                4 = Engineering   etc.)
                5 = Hotel Administration   9 = Graduate School

☐ 3. Class: 1 = Fresh.   2 = Soph.   3 = Junior   4 = Senior   5 = Graduate   6 = Other

☐ 4. Approximate grade in this course to date: 1 = A   3 = C   5 = F   6 = S
     (Note: S = C– or better)   2 = B   4 = D   7 = U

☐ 5. Approximate cumulative average   1 = 4.0   4 = 2.5   7 = 1.0
                                            2 = 3.5   5 = 2.0   8 = Not applicable
                                            3 = 3.0   6 = 1.5

☐ 6. Is this course in your intended or actual major?   1 = Yes   2 = No   3 = Undecided

☐ 7. My *most* important reason for taking this course was:
        1 = It is required for the major.   4 = It is required for graduate work.
        2 = It has a great reputation.   5 = Other
        3 = The subject matter was of interest.

Instructions: The following questions are to be answered using a 5-point scale, where "1" and "5" will be defined and "3" always stands for the midpoint. For example, if a course is slightly below the midpoint in a given aspect, mark a "2" for that item. Make *no* mark for questions that do not apply to this course.

☐ 8. The amount of work required by this course is appropriate for the credit received:
        1 = much less than appropriate   5 = much more than appropriate

☐ 9. Did the teacher stimulate your interest in the subject?
        1 = destroyed interest; was boring   5 = stimulated great interest

☐10. Was the teacher's presentation of material organized?
        1 = congested; disorganized   5 = clear; organized

☐11. Was the teacher willing to provide help for students who needed it?
        1 = seemed unwilling to help   5 = seemed interested in being helpful

☐12 Did the stated objectives of the course correspond with the outcome?
        1 = no agreement between announced objectives and what was taught

5 = considerable agreement between announced objectives and what was taught

☐ 13. Did the course offer opportunities to become familiar with the material through practice, discussion, or application to problems?
1 = no, few or no opportunities   5 = yes, many opportunities

☐ 14. The teaching skills of the teacher in this course, in comparison to my other teachers:
1 = much poorer than the majority   5 = much better than the majority

☐ 15. The value of this course to my overall education, in comparison to other courses:
1 = much less than from other courses   5 = much more than from other courses

☐ 16. The methods used to evaluate my knowledge and understanding of course material were:
1 = inadequate to assess my knowledge   5 = adequate to assess my knowledge

☐ 17. My opinion of this course is:
1 = very poor course   5 = an excellent course

☐ 18. The instructor deserves an overall rating of:
1 = very poor instructor   5 = very good instructor

Readings

(19–24) How valuable were the following readings? (Teacher will specify readings)
1 = worthless   5 = valuable; I learned a great deal

☐ 19. Reading A   ☐ 22. Reading D

☐ 20. Reading B   ☐ 23. Reading E

☐ 21. Reading C   ☐ 24. Reading F

☐ 25. How would you rate the amount of reading required for the course?
1 = much too light   5 = much too heavy

☐ 26. In general, how much overlap was there between the readings and the lectures?
1 = not enough overlap   5 = lectures repeated readings to an unnecessary degree

Papers and Project

☐ 27. Overall, how much did the assigned papers and project add to the value of the course?
1 = nothing; a useless exercise   5 = a great deal; I learned from the work

☐ 28. Were the criticisms of the papers and project adequate?
1 = too little feedback   5 = very instructive

☐ 29. Was the grading of papers and project fair?   1 = very unfair   5 = very fair

☐ 30. How would you rate the number of papers required?   1 = too few   5 = too many

Fieldwork

☐ 31. Was the fieldwork interesting?   1 = boring   5 = very interesting

☐ 32. Was the field instructor willing to help students who had difficulty?
1 = seemed unwilling to help   5 = seemed interested in being helpful

☐ 33. Was the relationship between lectures and fieldwork meaningful?
1 = no relationship   5 = the lectures and labs were well integrated

☐ 34. Overall, how would you rate the field instructor?   1 = very poor   5 = excellent

☐ 35. Overall, how much did you learn from the fieldwork?   1 = nothing   5 = a great deal

Examinations

☐ 36. Did the examinations adequately sample the important material in the course?
1 = not at all   5 = exam questions reflected the important aspects of the course

☐ 37. What was the nature of the exam items?
1 = too specific and detailed; picky   5 = too broad; easy to answer without facts

☐ 38. Did the exams make you think?   1 = not at all   5 = a great deal

☐ 39. Were the exams an interesting learning experience?   1 = not at all   5 = very definitely

☐ 40. How would you rate the length of the exams?   1 = not enough time given   5 = ample time

☐ 41. Were the exams free from unnecessary ambiguity?   1 = mostly ambiguous   5 = quite clear

☐ 42. How would you rate the difficulty of the exams?   1 = too easy   5 = too difficult

☐ 43. Was the type of examination (exercise, etc.) suitable for the purpose of the course?
1 = not at all   5 = very suitable for the purpose of the course

☐ 44. Was the grading of examinations fair?   1 = very unfair   5 = very fair

☐ 45. Was there adequate feedback as to what was expected on the exams?
1 = no answers or guidance given   5 = explanation of answers was provided

☐ 46. Overall, how would you rate the examinations in this course?
1 = very inadequate   5 = very adequate as a test of my knowledge

Appendix
Conversion factors
for U.S. and metric units

To Convert Column 1 into Column 2, Multiply by:	Column 1	Column 2	To Convert Column 2 into Column 1, Multiply by:
	Length		
0.621	kilometer, km	mile, mi	1.609
1.094	meter, m	yard, yd	0.914
0.394	centimeter, cm	inch, in	2.54
	Area		
0.386	kilometer², km²	mile², mi²	2.590
247.1	kilometer², km²	acre, acre	0.00405
2.471	hectare, ha	acre, acre	0.405
	Volume		
0.00973	meter³, m³	acre-inch	102.8
3.532	hectoliter, hl	cubic foot, ft³	0.2832
2.838	hectoliter, hl	bushel, bu	0.352
0.0284	liter	bushel, bu	35.24
1.057	liter	quart (liquid), qt	0.946
	Mass		
1.102	ton (metric)	ton (U.S.)	0.9072
2.205	quintal, q	hundredweight, cwt (short)	0.454
2.205	kilogram, kg	pound, lb	0.454
0.035	gram, g	ounce (avdp), oz	28.35
	Yield or Rate		
0.446	ton (metric)/hectare	ton (U.S.)/acre	2.24
0.892	kg/ha	lb/acre	1.12
0.892	quintal/hectare	hundredweight/acre	1.12
	Temperature		
$9/5\,(^{\circ}C + 32)$	Celsius	Fahrenheit	$5/9\,(^{\circ}F - 32)$
	$-17.8^{\circ}C$	$0^{\circ}F$	
	$0^{\circ}C$	$32^{\circ}F$	
	$100^{\circ}C$	$212^{\circ}F$	